4th International Workshop on Statistical Physics and Mathematics for Complex Systems

(SPMCS 2014)

Journal of Physics: Conference Series Volume 604

Yichang, China
12 – 16 October 2014

Editors:

A. Wang
S. Abe
W. Li

ISBN: 978-1-5108-0264-3
ISSN: 1742-6588

Printed from e-media with permission by:

Curran Associates, Inc.
57 Morehouse Lane
Red Hook, NY 12571

Some format issues inherent in the e-media version may also appear in this print version.

Printed by Curran Associates, Inc. (2015)

For permission requests, please contact the Institute of Physics
at the address below.

Institute of Physics
Dirac House, Temple Back
Bristol BS1 6BE UK

Phone: 44 1 17 929 7481
Fax: 44 1 17 920 0979

techtracking@iop.org

Additional copies of this publication are available from:

Curran Associates, Inc.
57 Morehouse Lane
Red Hook, NY 12571 USA
Phone: 845-758-0400
Fax: 845-758-2634
Email: curran@proceedings.com
Web: www.proceedings.com

TABLE OF CONTENTS

This volume contains 24 contributed papers presented at the 4th *International Workshop on Statistical Physics and Mathematics for Complex Systems* (*SPMCS*) held during October 12-16, 2014 in Yichang, China. Each paper was peer-reviewed by at least one referee chosen from a distinguished international panel.

The previous three workshops of this series were organized in 2008, 2010, and 2012, in Le Mans, France, Wuhan, China, and Kazan, Russia, respectively.

The SPMCS international workshop series is destined mainly to communicate and exchange research results and information on the fundamental challenges and questions in the vanguard of statistical physics, thermodynamics and mathematics for complex systems. More specifically, the topics of interest touch, but are not limited to, the following:

- Fundamental aspects in the application of statistical physics and thermodynamics to complex systems and their modeling
- Finite size and non-extensive system
- Fluctuation theorems and equalities, quantum thermodynamics
- Variational principle for random dynamics
- Fractal geometry, fractional mathematics

More than 50 participants from 7 countries participated in SPMCS-2014. 35 oral contributions were presented at the workshop.

We would like to take this opportunity to thank the members of the Scientific Program Committee, many of whom acted as reviewers of the papers and responded promptly. We would also like to thank the organizing committee, the session chairs, the technicians and the students for the smooth running of the whole workshop. Thanks also go to China Three Gorges University who provided generous support for the conference venue, as well as exquisite refreshments for the tea breaks. The workshop was also partially supported by Central China Normal University and the Programme of Introducing Talents of Discipline to Universities under grant NO. B08033. Special thanks are due to Ms Juy Zhu who has done excellent editing work with great effort.

Editors: Alexandre Wang, Sumiyoshi Abe, and Wei Li

Peer review statement

All papers published in this volume of *Journal of Physics: Conference Series* have been peer reviewed through processes administered by the proceedings Editors. Reviews were conducted by expert referees to the professional and scientific standards expected of a proceedings journal published by IOP Publishing.

On the variational principle for fractional kinetic theory

Sumiyoshi Abe[1,2], **Akifumi Oohata**[1]

[1] Department of Physical Engineering, Mie University, Mie 514-8507, Japan
[2] Institute of Physics, Kazan federal University, Kazan 420008, Russia

E-mail: suabe@sf6.so-net.ne.jp (S.Abe)

Abstract. In a recent paper (Abe S 2013 *Phys. Rev.* E **88** 022142), a variational principle has been formulated for spatiotemporally-fractional Fokker-Planck equations and applied to derivations of their approximate analytic solutions based on the Lévy Ansatz. Here, the problem of the constraint associated with normalization condition on a probability distribution behind the principle is discussed. It is shown that the action functional possesses a specific transformation property in terms of an auxiliary field and the constraint turns out to have already been imposed implicitly in terms of such a structure.

1. Introduction

Variational principle has been playing an outstanding role in science and will be doing so over the future. It offers a unified approach to deriving basic equations as stationarity conditions on relevant functionals. In physics, functionals are often actions that are given in terms of Lagrangians or Hamiltonians. Therefore, time-reversal invariance is essential, leading to an obvious problem for systems without such an invariance, e.g., dissipative systems. This point can, however, be overcome if the space of variables is extended.

In a recent work [1], a variational principle has been formulated for spatiotemporally-fractional Fokker-Planck equations in the context of anomalous diffusion and applied to systems of physical interest as a method for obtaining approximate analytic solutions with the help of the Lévy Ansatz. There, an auxiliary field has been introduced in order to extend the space of variables because of the absence of time-reversal invariance in the equations. However, an issue regarding the constraint associated with the normalization condition on a probability distribution has not been discussed in that work. The purpose of this short note is to clarify this point in connection with a specific transformation property of the action.

2. Fractional Fokker-Planck equation

The fractional Fokker-Planck equation discussed in Ref. [1] is of the following form:

$$\frac{\partial p(x,t)}{\partial t} = {}_0D_t^{1-\alpha}\left\{-\frac{\partial}{\partial x}[F(x)p(x,t)] - D^*(-\Delta)^{\gamma/2}p(x,t)\right\}. \tag{1}$$

Here, $p(x,t)dx$ is the probability of finding a particle (a random walker) in the interval $[x, x+dx]$ at time t. The domain of $p(x,t)$ is $(-\infty,\infty)\times[0,T]$. D^* is a generalized diffusion coefficient, and the first term inside the braces on the right-hand side is a drift. ${}_0D_t^{1-\alpha}$ and $-(-\Delta)^{\gamma}/2$ are the Riemann-Liouville fractional differential operator and the Riesz fractional " Laplacian " [2,3], respectively. The ranges of α and γ of physical interest are

$$0 < \alpha < 1, \qquad 0 < \gamma < 2. \tag{2}$$

Then, the Riemann-Liouville operators are defined by

$${}_0D_t^{1-\alpha} f(t) = \frac{1}{\Gamma(\alpha)} \frac{d}{dt} \int_0^t ds (t-s)^{\alpha-1} f(s), \tag{3}$$

$${}_tD_T^{1-\alpha} f(t) = \frac{1}{\Gamma(\alpha)} \left(-\frac{d}{dt}\right) \int_t^T ds (s-t)^{\alpha-1} f(s), \tag{4}$$

which fulfill the following relation:

$$\int_0^T dt [{}_0D_t^{1-\alpha} f(t)] g(t) = \int_0^T dt f(t) [{}_tD_T^{1-\alpha} g(t)]. \tag{5}$$

On the other hand, the Riesz operator is defined by

$$-(-\Delta)^{\gamma/2} = -\frac{1}{2\cos(\pi\gamma/2)} \left[\frac{d^\gamma}{dx^\gamma} + \frac{d^\gamma}{d(-x)^\gamma}\right], \tag{6}$$

which satisfies

$$\int_{-\infty}^{\infty} dx [-(-\Delta)^{\gamma/2} \psi(x)] \phi(x) = \int_{-\infty}^{\infty} dx \psi(x) [-(-\Delta)^{\gamma/2} \phi(x)]. \tag{7}$$

It is noted that the Riesz differentiation of a constant vanishes, whereas the Riemann-Liouville differentiation does not.

The physical mechanism underlying Eq. (1) is described in the language of continuous-time random walks [4] as follows. Temporal fractionality characterized by α implies that the waiting-time distribution decays as a power law, $P_W(t) \sim t^{-1-\alpha}$, and does the jump distribution too, $P_J(x) \sim |x|^{-1-\gamma}$. The power-law waiting (jumping) suppresses (enhances) diffusion. Therefore, resulting the diffusion property is characterized by the interplay between these two effects. In particular, presence of long jumps has motivated the work in Ref. [1] to propose the Lévy Ansatz. There, a periodic drift is considered as an example and, with the help of the Rayleigh-Ritz-like method, it is shown that motion of the center of the Lévy distribution is slow (i.e., not an exponential-law behavior familiar in ordinary kinetic theory but a power-law one) and the diffusion property is characterized by $l \sim t^{\alpha/\gamma}$, where l stands for the spatial extension of the distribution such as the half width. A particular case with $\alpha/\gamma = 1/2$ is an example of the so-called non-Gaussian normal diffusion.

3. Action and variational principle

Clearly, the fractional Fokker-Planck equation (1) is not invariant under time reversal. To construct the corresponding action functional, an auxiliary field $\Lambda(x,t)$ has been introduced in Ref. [1] in order to extend the space of variables. It is noted that $\Lambda(x,t)$ is not a probability distribution. The action presented reads

$$I[p, \Lambda] = -\int_0^T dt < \dot{\Lambda} + {}_tD_T^{1-\alpha} \{F\Lambda' - D^*(-\Delta)^{\gamma/2}\Lambda\} > - < \Lambda > |_{t=0}, \tag{8}$$

where $\dot{\Lambda} = \partial\Lambda/\partial t$, $\Lambda' = \partial\Lambda/\partial x$ and $< A >= \int_{-\infty}^{\infty} dx A(x,t)p(x,t)$. Introduction of the last term on the right-hand side is inspired by the works in Refs. [5,6], where a variational principle for the Liouville-von Neumann equation is discussed (note that the Liouville-von Neumann equation possesses time-reversal invariance in contrast to the present fractional Fokker-Planck equation).

The variations of this action with respect to Λ and p are found to be given as follows:

$$\delta_\Lambda I = \int_0^T dt \int_{-\infty}^{\infty} dx \{\dot{p} + {}_0D_t^{1-\alpha}[(Fp)' + D^*(-\Delta)^{\gamma/2}p]\}\delta\Lambda - \int_{-\infty}^{\infty} dx\ p\delta\Lambda\Big|_{t=T}, \tag{9}$$

$$\delta_p I = -\int_0^T dt \int_{-\infty}^{\infty} dx \{\dot{\Lambda} + {}_tD_T^{1-\alpha}[F\Lambda' - D^*(-\Delta)^{\gamma/2}\Lambda]\}\delta p - \int_{-\infty}^{\infty} dx\ \Lambda\delta p\Big|_{t=0}, \tag{10}$$

where the relations in Eqs. (5) and (7) have been used. The temporal boundary terms in these equations vanish if $\delta\Lambda(x,T) = 0$ and $\delta p(x,0) = 0$. To realize them, let us impose the following temporal boundary conditions:

$$\Lambda(x,T) = \Lambda_0(x), \tag{11}$$

$$p(x,0) = \delta(x - X_0), \tag{12}$$

where $\Lambda_0(x)$ is a fixed function and X_0 is a constant. Actually, $p(x,0)$ in Eq. (12) can be replaced by an arbitrary normalized distribution, but this point does not affect the subsequent discussion.

In Ref. [1], it is claimed that the stationarity conditions, $\delta_\Lambda I = 0$ and $\delta_p I = 0$, lead to Eq. (1) and

$$-\frac{\partial\Lambda(x,t)}{\partial t} = {}_tD_T^{1-\alpha}\left\{F(x)\frac{\partial\Lambda(x,t)}{\partial x} - D^*(-\Delta)^{\gamma/2}\Lambda(x,t)\right\}, \tag{13}$$

respectively.

It is of interest to observe that the field equation (13) has a structure similar to Eq. (1) with time reversal. In particular, the fractional operator ${}_tD_T^{1-\alpha}$, which is the adjoint of ${}_0D_t^{1-\alpha}$ in Eq. (1), appears through the relation in Eq. (5). This can be thought of as the invariance in the present system.

4. Transformation property of the action and normalization constraint

Now, the problem in the above scheme is that the variation with respect to the probability distribution should be performed under the constraint associated with the normalization condition

$$\int_{-\infty}^{\infty} dx\ p(x,t) - 1 = 0, \tag{14}$$

which is preserved in time by Eq. (1). Below, we shall see that this problem is indivisibly related to a specific transformation property that the action possesses.

A key point is in the fact that the auxiliary field Λ is not a probability distribution and does not have to be normalized. [Even if Λ is forced to be normalized at certain time, the field equation (13) does not preserve such a condition.] This allows us to transform the field. The transformation we consider here is the following one:

$$\Lambda(x,t) \to \Lambda(x,t) + \int_t^T ds\lambda(s). \tag{15}$$

It is important to note that this keeps the final condition in Eq. (11) unchanged. Under this, the action in Eq. (8) is found to transform as follows:

$$I[p,\Lambda] \to I[p,\Lambda] + \int_0^T dt\lambda(t)\left[\int_{-\infty}^{\infty} dx p(x,t) - 1\right], \tag{16}$$

where Eq. (12) has been used. Identifying λ with a Lagrange multiplier, we see that the second term is precisely the constraint associated with the normalization condition in Eq. (14). In other words, even if the constraint would be included in the action, it could be eliminated by the inverse transformation of Eq. (15). Therefore, the constraint turns out to have already been imposed implicitly in the action (8) through additive arbitrariness of Λ, which is somewhat analogous to the gauge-theoretic structure. This justifies the scheme in Section 3.

5. Concluding remarks

We have reexamined the variational principle that has recently been developed for spatiotemporally-fractional Fokker-Planck equations. In particular, we have discussed the problem of the constraint associated with the normalization condition behind the principle. We have shown that the action possesses a specific transformation property concerning additive arbitrariness of the auxiliary field and the constraint has already been imposed implicitly.

In Ref. [6], it is pointed out for the variational principle for the Liouville-von Neumann equation that the constraint of the normalization condition on a density matrix can be taken care of if the phase of an auxiliary matrix is redefined. This idea may be analogous to ours expressed in Eq. (15).

Besides its useful role in obtaining approximate analytic solutions of kinetic equations, a variational principle also provides a powerful theoretical framework for discussing symmetry of probability distributions. It is shown e.g. in Refs. [7,8] how power-law distributions as the solutions of a class of non-fractional Fokker-Planck equations can be related to dilatation symmetry [note that in those papers, the temporal boundary terms like in Eq. (8) are not contained]. It is of interest to develop such a symmetry discussion for fractional kinetics.

Acknowledgments

One of the authors (S. A.) would like to thank the organizers of The 4th International Workshop on Statistical Physics and Mathematics for Complex Systems (12-16 October, 2014, Yichang, China) for inviting him to give a lecture. His work has been supported in part by a Grant-in-Aid for Scientific Research from the Japan Society for the Promotion of Science and by the Ministry of Education and Science of the Russian Federation (federal program of competitive growth of Kazan Federal University).

References

[1] Abe S 2013 *Phys. Rev.* E **88** 022142
[2] Hilfer R (ed) 2000 *Applications of Fractional Calculus in Physics* (Singapore: World Scientific)
[3] Herrmann R 2014 *Fractional Calculus: An Introduction for Physicists* 2nd ed (Singapore: World Scientific)
[4] Metzler R and Klafter J 2000 *Phys. Rep.* **339** 1
[5] Balian R and Vénéroni M 1981 *Phys. Rev. Lett.* **47** 1353 ; 1981 *ibid.* **47** 1765(E)
[6] Éboli O, Jackiw R and Pi S Y 1988 *Phys. Rev.* D **37** 3557
[7] Abe S 2004 *Phys. Rev.* E **69** 016102
[8] Abe S 2007 *Complexity, Metastability and Nonextensivity* ed Abe S, Herrmann H, Quarati P, Rapisarda A and Tsallis C, AIP Conference Proceedings **965**

SPMCS2014 IOP Publishing
Journal of Physics: Conference Series **604** (2015) 012002 doi:10.1088/1742-6596/604/1/012002

ROLE OF CLOSED PATHS IN THE PATH INTEGRAL APPROACH OF STATISTICAL THERMODYNAMICS.

Jean-Pierre Badiali

Université Pierre et Marie Curie 4 Place Jussieu, 75230 Paris Cedex 05, France
Institute for Condensed Matter Physics of National Academy of Sciences of Ukraine,Lviv, Ukraine.

E-mail: `jpbadiali@numericable.com`

Abstract. Thermodynamics is independent of a description at a microscopic level consequently statistical thermodynamics must produce results independent of the coordinate system used to describe the particles and their interactions. In the path integral formalism the equilibrium properties are calculated by using closed paths and an euclidean coordinate system. We show that the calculations on these paths are coordinates independent. In the change of coordinate systems we consider those preserving the physics on which we focus. Recently it has been shown that the path integral formalism can be built from the real motion of particles. We consider the change of coordinates for which the equations of motion are unchanged. Thus we have to deal with the canonical transformations. The Lagrangian is not uniquely defined and a change of coordinates introduces in hamiltonians the partial time derivative of an arbitrary function. We have show that the closed paths does not contain any arbitrary ingredients. This proof is inspired by a method used in gauge theory. Closed paths appear as the keystone on which we may describe the equilibrium states in statistical thermodynamics.

1. Introduction

The path integral method has been proposed by Feynman [1] as an alternative to the Schrödinger equation. Today it is one theoretical tool among the most used in quantum field theory [2]. The path integral formalism or the functional integral point of view has been extended in statistical thermodynamics (see for instance [3]). In quantum physics this formalism has been extensively used but due to the presence of a complex measure it is difficult to have rigorous mathematical treatments. In contrast in statistical physics the measure is positive and the functional analysis has a rigorous mathematical basis [2].

In the initial version of statistical physics the path integral formalism requires to solve the Schödinger equation, to use the canonical form of the density matrix and to introduce some mathematical tricks. Thus all the machinery of quantum mechanics is needed to calculate thermodynamic quantities. Feynman ([1] see " Remarks on methods of derivation" p. 295) suggested that it must be possible to calculate more directly the thermodynamic quantities without solving the Schrödinger equation but directly starting from the time-dependent motion of particles. In a recent paper [4] it has been shown that it is possible to follow this route. The trajectories of particles are quantified by the path integral formalism, only the closed paths

have to be inspected on a time associated with a non usual equilibrium condition. In such an approach the time is a real time and the path integral is not just a formal mathematical trick. One interest of this route is to offer a scheme in which thermodynamic equilibrium and irreversible processes can be treated on the same footing [4]. Schrödinger equation and density matrix are not used. It is interesting to note that the thermodynamics of black holes can be obtained without using all the machinery of the Einstein equations [5]. In this paper we want to show that the thermodynamic properties usually calculated in an euclidean coordinate system are in fact independent of the coordinate system.
In Section 2 we briefly summarize an approach recently proposed in agreement with the Feynman remarks [4]. In Section 3 we use a general coordinate system and introduce a traditional result associated with canonical transformations. In Section 4 we develop an approach reminiscent of the one proposed by Fadeev and Popov [6] for gauge theories. A short conclusion is given in Section 5.

2. Thermodynamic quantities in terms of path integrals

Feynman [3] observed that to the density matrix ρ we may associate a quantity $\rho(u) = \exp \frac{-Hu}{\hbar}$ where u has been redefined to be $\beta\hbar$; $\beta = \frac{1}{k_B T}$ in which T is the temperature. This quantity verifies the differential equation

$$\hbar \frac{\partial \rho(u)}{\partial u} = -H\rho(u) \tag{1}$$

that is formally a Schrödinger equation in which we have made the transformation $it = u$. This remark implies that the partition function can be expressed in term of a path integral as it has been done for the wave function. The partition function Q can be written [3]

$$Q = \frac{1}{N!} \int dx^N(0) \int \mathcal{D}x^N(t) exp(-\frac{1}{\hbar} \int_0^{\beta\hbar} H(s)ds) \tag{2}$$

where $x^N(0)$ represents the set of the positions $x^i(0)$ occupied by the N particles at the time $t = 0$ and $x^N(t)$ is a similar quantity but associated with the time t, $\mathcal{D}x^N(t)$ is the path integral measure. The calculation of path integrals has been presented in many textbooks (see for instance [7], [8]). In order to calculate Q we introduce a time-discretization having $\delta t = \frac{\beta\hbar}{n}$ as time step, n goes to infinity and δx the difference of position corresponding to δt. The path integral measure is given by

$$\mathcal{D}x^N(t) = \frac{1}{C} \prod_{i=1}^{N} \prod_{j=1}^{n-1} dx_j^i \tag{3}$$

in which C is a normalization constant and x_j^i represents the position of the particle i at the time $j\delta t$, we have $x_0^i = x^i(0)$ and $x_{n-1}^i = x^i(t - \delta t)$. In (2) we must take $x^i(0) = x^i(\beta\hbar)$ showing that we only consider closed paths, this is associated with the fact that in traditional version of statistical mechanics we only focus on the trace of the density matrix. The Hamiltonian is written $H(s) = K(s) + U(s)$ where $K(s)$ is given by

$$K(s) = \sum_{i=1}^{N} \frac{1}{2} m (\frac{dx^i(s)}{ds})^2 = \sum_{i=1}^{N} K^i(s) \tag{4}$$

and

$$U(s) = \sum_{i=1}^{N} U^i(s) \tag{5}$$

where $U^i(s)$ is the total potential acting on the particle i located at the point x^i at the time s. Written in the form (2) the calculation is performed as follows. For each particle i we focus on

its position $x^i(0)$ fixed at the time 0 and we consider all the closed paths formed from $x^i(0)$ and finally we integrate on all the value of $x^i(0)$.
Using the relation $F = -k_B T \ln Q$ between the free energy F and Q and the thermodynamic relation $F = U - TS$ we obtain in terms of path integral

$$S = k_B \ln \frac{1}{N!} \int dx^N(0) \int \mathcal{D}x^N(t) exp(-\frac{1}{\hbar} \int_0^{\beta\hbar} [H(s) - U] ds) \tag{6}$$

showing that the entropy is determined by the fluctuations of the internal energy along the paths these fluctuations being such as $\int_0^{\tau} (H(s) - U) ds \approx h$. Instead of the Boltzmann formula in which we count a number of states here we count a number of paths.
The path integral formalism developed above corresponds to the initial traditional route. In [4] we started on the investigation of a real motion in classical space-time [4]. We count the closed paths explored during a time τ. If we require that τ must be such the average of energy counted on the paths is identical to the free energy needed to create the system we find that $\tau = \beta\hbar$. The time τ is a characteristic of equilibrium in no way it must be considered as the relaxation time characterizing how a system in non-equilibrium relaxes toward an equilibrium state. τ represents the natural unit of time and it introduces a problem of measurement for the thermodynamic properties [4]. If a measurement is performed on a time interval smaller than τ the result will be unpredictable due to the quantum fluctuations. With this point of view all the equilibrium properties are calculated from a dynamic point of view but the results are identical to those deduced via the Gibbs ensemble method.
To summarize, in this approach the time we consider is the usual time *i.e.* a real quantity and we have to deal with the usual dynamics. The paths represent the particles trajectories that are quantified via the functional integration in which the Heisenberg uncertainty relations appear. With this approach we do not use the Schrödinger equation, this appears as a necessity if we want to use the same formalism for describing both equilibrium states and time-irreversible processes. Another important point of this approach is to show that the passage from statistical mechanics to quantum physics can be analyed in term of time-irreversibility.

3. The canonical transformations

The Hamiltonian $H(s)$ introduced in (2) and defined by (4) and (5) is given in an euclidean coordinate system. The thermodynamic quantities ignoring a microscopic description must be coordinates independent. Thus we have to rewrite $H(s)$ in a general coordinate system. Of course among all the coordinates system we must keep those preserving the investigated physics. We decide to inspect all the coordinate systems verifying the same equation of motion. This leads to focus on the canonical transformations [9].
To save the notations we consider a system reduced to one particle, the generalization is straightforward. The kinetic energy (4) is now reduced to $K(x, s) = \frac{1}{2} m (\frac{dx(s)}{ds})^2$ and the potential $U(x, s)$ is the external potential. Now we restrict x to represent an euclidean coordinate system and we use the symbol q for a general coordinate system. Basically the motion is described via the optimization of a Lagrangian giving rise to the Euler equations. However two lagrangians differing by the total time derivative of an arbitrary function depending on time and position give to the same motion. Let consider two general coordinate systems for which the positions are referred by q and q'. Then we have for the change in hamiltonians [9]

$$H'(q', t)) = H(q, t) + \frac{\partial F(q, t)}{\partial t} \tag{7}$$

in which $F(q, t)$ is an arbitrary function. Now we have to calculate on the closed paths

$$\int_0^{\tau} H(q', t) dt = \int_0^{\tau} [H(q, t) + \frac{\partial F(q, t)}{\partial t}] dt \tag{8}$$

The integral on $\frac{\partial F(q,t)}{\partial t}$ can be performed easily, we have

$$\int_0^\tau \frac{\partial F(q',t)}{\partial t}dt = \delta t \sum_{j=1}^{n} \frac{F(q_j,t_j) - F(q_{j-1},t_{j-1})}{\delta t} = F(q_0,\tau) - F(q_0,0) \tag{9}$$

Thus the integration on closed paths starting from q_0 introduces to a factor $\Lambda(q_0) = \frac{1}{\hbar}[F(q_0,\tau) - F(q_0,0)]$. To derive this result we have taken into account that the path integral requires a discretization but in this case the result is identical to the one obtained by considering continuous variables.
Now we can consider two results obtained with two different coordinates one referred by q' and the second by q. We have for (2) in the case $N = 1$

$$Q' = \int dq'(0) \int \mathcal{D}q'(t) exp(-\frac{1}{\hbar}\int_0^\tau H(q',t)dt) \tag{10}$$

$$Q = \int dq(0) \int \mathcal{D}q(t) exp(-\frac{1}{\hbar}\int_0^\tau H(q,t)dt) \tag{11}$$

It is well know [10] that functional measure is invariant in a translation of the variables producing a change of coordinates and hence $dq_0\mathcal{D}q(t) = dq'_0\mathcal{D}q'(t)$ but with this result and (8) we transform (10) as

$$Q' = \int dq(0) \int \mathcal{D}q(t) exp(-\frac{1}{\hbar}\Lambda(q_0)\int_0^\tau H(q,t)dt) \tag{12}$$

from which we can see that Q and Q' are different. The closed paths induce a monodromy defined by $\Lambda(q_0)$. We have to deal with something similar to a gauge theory (see for instance [10]): the basic equations give the same physics but they contain an arbitrary function $F(q,t)$ creating different expressions for the same quantities. Since the physical results must be independent of $F(q,t)$ it must exist a degree of freedom from which the we may eliminate it. Due to the similitude with the gauge theory it seems suited to introduce a method efficient in quantum field theory [6].

4. Invariance on the coordinate systems

In standard quantum mechanics the expectation $\langle A\rangle$ of a given operator A is calculated via the density matrix ρ according to $\langle A\rangle = Tr(\rho A)$ where $Tr(B)$ means that we have to take the trace of the operator B. To calculate $Tr(\rho A)$ we need to introduce a complete basis of orthogonal and normalized vectors. It is very well known that the trace is independent of the basis used, to demonstrate this we use the closure relation $\sum_i |i><i| = 1$ that we can consider as a decomposition of the unity. In addition, in the demonstration of the trace independence we perform a change in the order of the summation. Similar arguments will be used to show the independence of the thermodynamic variables on the choice of the coordinate system. However these arguments are implemented by a method used in gauge theory [6].
Here the decomposition of the unity wil be build via the Dirac distribution defined according to

$$\int \delta(y-x)f(x)dx = f(y) \tag{13}$$

For each value of q_0 we have associated a number $\Lambda(q_0)$ depending on the arbitrary function $F(q,t)$. By changing, for instance, some parameters involved in the definition of $F(q,t)$ we can change its value and we assume that it can reach a value $\kappa(q_0)$ where $\kappa(q_0)$ is a given function of q_0. Using (13) for the function $f(x) = 1$ we get

$$\int \delta(\kappa(q_0) - \Lambda(q_0))d\Lambda(q_0) = 1 \tag{14}$$

If (14) is inserted in (12) the result is unchanged and independent of $\kappa(q_0)$ and we can perform an integration on a normalized function $g(\kappa(q_0))$. Now we have

$$Q' = \int dq_0 d\kappa(q_0) g(\kappa(q_0)) d\Lambda(q_0) \delta(\kappa(q_0) - \Lambda(q_0)) \exp -\frac{1}{\hbar}\Lambda(q_0) \int \mathcal{D}q(t) exp(-\int_0^{\mathcal{T}} H(q,t)dt) \quad (15)$$

If we perform first an integration over $\kappa(q_0)$ we get

$$Q' = Cte \int dq_0 \int \mathcal{D}q(t) exp(-\int_0^{\mathcal{T}} H(q,t)dt) \quad (16)$$

in which the constant Cte is given by

$$Cte = \int d\Lambda(q_0) g(\Lambda(q_0)) \exp -\frac{1}{\hbar}\Lambda(q_0) \quad (17)$$

The function $g(\kappa(q_0))$ which indicates the distribution of $\kappa(q_0)$ is chosen independently of $\Lambda(q_0)$ and consequently the constant Cte is independent of $\Lambda(q_0)$. For instance, to be illustrative we can choose for $g(x)$ a Gaussian distribution

$$g(x) = (\frac{A}{\pi})^{\frac{1}{2}} \exp Ax^2 \quad (18)$$

The value of Cte is $\exp \frac{1}{4\hbar^2 A}$. Since the dimension of $\Lambda(q_0)$ is the same as $\hbar$ we can take $A = \frac{1}{\hbar^2}$ and finally Cte is just a numerical factor $C = \exp \frac{1}{4}$.

5. Conclusion

We can say that the exact definition of the partition function is given by (16), it differs from the traditional one by a numerical factor. But the main result is that the thermodynamic properties are independent of the coordinate system. This is the consequence of the liberty we have to choose the function $g(x)$ which is independent of the arbitrary function $F(q,t)$. One choice of $g(x)$ has been proposed to be illustrative. Our demonstration is inspired by the properties of the trace of the density matrix and by the Fadeev-Popov method [6] used in gauge theory. Finally, we may conclude that the closed paths are strongly associated with equilibrium thermodynamics.

References

[1] Feynman R.P. and Hibbs A.R. 1965 *Quantum mechanics and path integrals* (New York: Mc Graw Hill)
[2] Glimm J. and Jaffe A. 1987 *Quantum Physics. A functional Integral Point of View* (New York: Springer Verlag)
[3] Feynman R.P. 1972 *Statistical mechanics. A set of lectures* (Frontiers in Physics) (New York: Benjamin)
[4] Badiali J.P. 2013 arXiv:1311.4995v1 [cond-math.stat-mech]
[5] Wald R.M 1999 *Class. Quantum Grav.* **16** 77
[6] Fadeev L.D. and Popov V.N. 1967 *Phys. Lett.* B **25** 29
[7] Schulmann L.S. 1996 *Techniques and applications of paths integrals* (New York: Wiley)
[8] Kleinert H 1995 *Path integrals in quantum mechanics, statistics and polymer physics* (Singapore:World Scientific)
[9] Arnold V.I. 1989 *Mathematical Methods in Classical Mechanics* (New York:Springer Verlag)
[10] Rivers R.J. 1990 *Path integral methods in quantum field theory* (Cambridge: Monographs on Mathematical Physics)

Journal of Physics: Conference Series **604** (2015) 012003 doi:10.1088/1742-6596/604/1/012003

A study of the electrical properties of complex resistor network based on NW model

Yunfeng Chang, Yunting Li, Liu Yang, Lu Guo and Gaochao Liu

College of Science, China Three Gorges University, Yichang, Hubei Province, China

E-mail: changyf_ctgu@163.com

Abstract. The power and resistance of two-port complex resistor network based on NW small world network model are studied in this paper. Mainly, we study the dependence of the network power and resistance on the degree of port vertices, the connection probability and the shortest distance. Qualitative analysis and a simplified formula for network resistance are given out. Finally, we define a branching parameter and give out its physical meaning in the analysis of complex resistor network.

1. Introduction

Complex networks have recently emerged as an invaluable tool for describing and quantifying complex systems in many branches of science[1, 2, 3, 4]. Suppose the edges are resistors and the vertices are junctions between resistors, and suppose we apply a voltage or current between two vertices s and t such that a current I flows from s to t through the network. What then is the property of complex network composed by resistors? In 1993, Klein proposed the concept of resistance distance of graphs[5]. In 2010, Newman gave a brief introduction on how to solve vertex voltage in active two-port complex resistor network[4]. Along with the in-depth study of the physical meaning and the topological feature of complex network, many complex networks are found to display community structure[6, 7, 8], dividing naturally into communities or modules with dense connections within communities but sparse connections among them. Communities have proven to be of interest both in their own right as functional building blocks within networks and for the insights they offer into the dynamics or modes of the formation of networks. A large volume of research has been devoted to the development of algorithmic tools for discovering communities[9, 10, 11, 12, 13]. In 2004, Wu and Huberman presented a fast spectrum segmentation method based on resistor network voltage spectrum, combining with the Kirchhoff's current law, the segmentation of community structure was solved in a better way[14].

For the explicit computation of two-port resistance, much work has been done on arbitrary finite and infinite lattices in terms of the eigenvalues and eigenvectors of the Laplacian matrix[15, 16, 17, 18]. That is, most studies of resistor network focus on the calculation of equivalent resistance of regular networks[19] while the study of resistor network based on complex network model is relatively less. In this paper, two-port resistor network based on NW small world complex network model[20] are studied. Mainly, we study the dependence of the network power and resistance on the degree of port vertices, the connection probability and the shortest distance. Qualitative analysis and a simplified formula for network resistance are given out.

Finally, we define a branching parameter and give out its physical meaning in the analysis of complex resistor network.

2. Electrical properties of resistor network

Suppose we apply a voltage or current between two vertices s and t such that there is a current I flows from s to t through the network, where s is the input vertex and t is the output vertex($s \neq t$). By setting the reference voltage of the output vertex t as $U_t = 0$, the relative voltage of the input vertex s as U_s, the electrical parameters of complex resistor network are given below.

2.1. The resistance of complex resistor network

In this paper, network resistance is divided into two types: network resistance of vertex-pair $R(s,t)$ and network resistance of vertex-degree $R(k_s, k_t)$. $R(s,t)$ is the equivalent total network resistance from the input vertex s to the output vertex t, such that:

$$R(s,t) = \frac{U_s}{I}. \tag{1}$$

As for fixed input-output vertex-pair (s,t), $R(s,t)$ is a fixed value independent of the applied voltage or current, we get $R(s,t) = R(t,s)$. Network resistance of vertex-degree $R(k_s,k_t)$ is defined as the average value of network resistance of vertex-pair with the same input-output degree (k_s, k_t):

$$R(k_s,k_t) = \frac{1}{N(k_s,k_t)} \sum_{s\in V_s, t\in V_t} R(s,t). \tag{2}$$

Where $N(k_s,k_t)$ is the number of vertex-pair whose degrees are (k_s,k_t), V_s is the collection of vertices whose degree is k_s, V_t is the collection of vertices whose degree is k_t. $R(k_s,k_t)$ is also independent of the applied voltage and current $R(k_t,k_s) = R(k_s,k_t)$.

2.2. The power of complex resistor network

Network power is also divided into two types: network power of vertex-pair $P(s,t)$ and network power of vertex-degree $P(k_s,k_t)$. $P(s,t)$ is the total power consumption of the complex resistor network when a current I is inputted at vertex s and flow out from vertex t:

$$P(s,t) = IU_s. \tag{3}$$

On account of that different vertex-pairs in the network may have the same degree as (k_s,k_t), the average value of the total power of these vertex-pairs is called the network power of vertex-degree, denoted by $P(k_s,k_t)$:

$$P(k_s,k_t) = \frac{1}{N(k_s,k_t)} \sum_{s\in V_s, t\in V_t} P(s,t). \tag{4}$$

Here $P(s,t) = P(t,s)$, $P(k_s,k_t) = P(k_t,k_s)$

3. The relationship between the electrical properties and the topological structure of complex resistor network

In general, small resistor network can be solved by its properties of series connection and parallel connection: the more the branches of the network have, the more parallel resistors it have and the smaller the total resistance will be, thus the network current streaming capability will be strong. For large resistor network based on a complex network, network voltage division capability and current streaming capability can be used to describe the affection of complex network topological structure on network electrical properties. In this paper, we select network power and resistance to describe network's voltage division capability and current streaming capability. For the same complex resistor network, when the input current is constant, the smaller the network resistance of vertex-pair $R(s,t)$, the smaller the power of vertex-pair $P(s,t)$ and the stronger the current streaming capability, otherwise the current streaming capability will be weak correspondingly; when the input voltage remains constant, the bigger the network resistance of vertex-pair $R(s,t)$, the smaller the network power of vertex-pair $P(s,t)$ and the stronger the voltage division capability, or it will be weak. As for different complex resistor network, we can evaluate the current streaming capability of the network by comparing the network resistance of vertex-degree $R(k_s,k_t)$ and the network power of vertex-degree $P(k_s,k_t)$.

Evidently, the electrical properties of complex resistor network have great relationship with its underlying topological structure. Even for the same network model, when the input and output vertices are different, the corresponding electrical parameters will be different. However, the statistical features such as degree distribution, shortest distance distribution, etc. will remain unchanged. Then, what's the relationship between these electrical parameters and the statistical features of the complex networks? We study this problem on the basis of NW small-world complex network model.

3.1. The Relationship between network power and input-output vertex degree

The resistor network here is a small world complex network with $N = 1000$ vertices based on NW small world complex network model. The unit of resistance and current was omitted for convenience. By setting all the edge resistance $R_0 = 1$, the input current $I = 10$, the connection probability $p = 0.002$(average degree $\langle k\rangle = 5.84$) and $p = 0.04$(average degree $\langle k\rangle = 80.53$), and the shortest distance from the input vertex to the output vertex $d = 1$, Figure 1-6 give out the relationship between network power $P(s,t), P(k_s,k_t)$ and the input-output degree (k_s,k_t). Figure 1 and Figure 2 are three dimensional evolution of the network power of vertex-pair $P(s,t)$ along with the input and output vertex degree k_s and k_t with different connection probabilities. Figure 3 and Figure 4 are the change of network power of vertex-pair $P(s,t)$ with the change of output vertex degree k_t while the input vertex degree k_s remains unchanged. Figure 5 and Figure 6 are the change of network power of vertex-degree $P(k_s,k_t)$ with the change of output vertex degree k_t while the input vertex degree k_s remains unchanged.

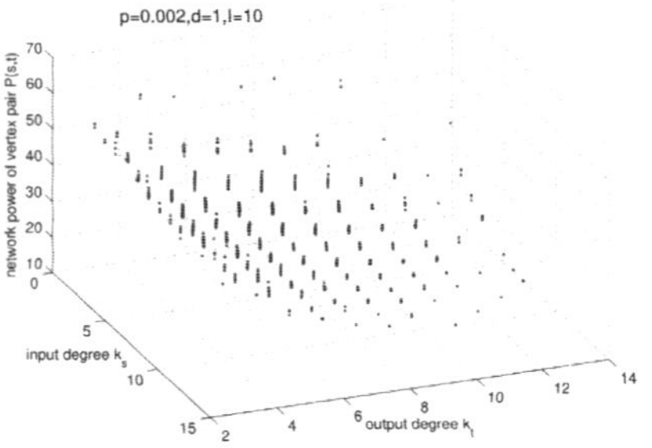

Figure 1. Three dimensional evolution of network power of vertex-pair $P(s,t)$ along with the input and output vertex degree k_s and k_t when the connection probability $p = 0.002(\langle k\rangle = 5.84)$.

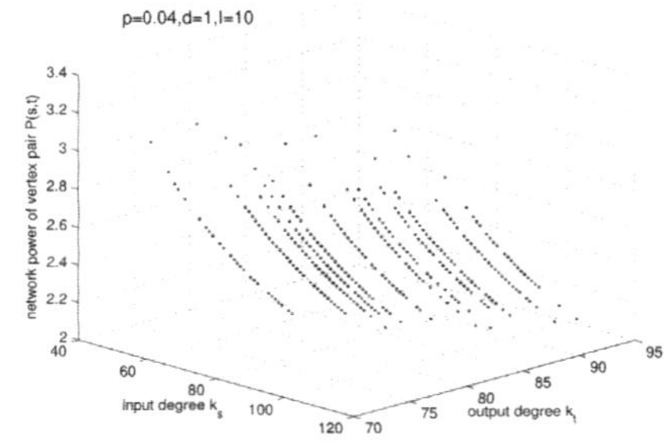

Figure 2. Three dimensional evolution of network power of vertex-pair $P(s,t)$ along with the input and output vertex degree k_s and k_t when the connection probability $p = 0.04(\langle k\rangle = 80.53)$.

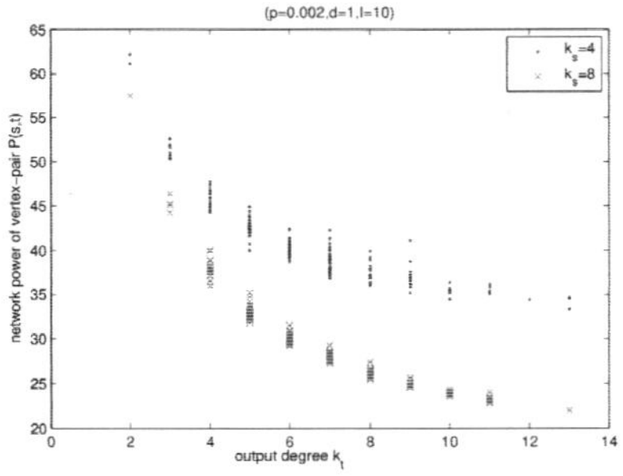

Figure 3. The change of network power of vertex-pair $P(s,t)$ with the change of output vertex degree while the input vertex degree remains unchanged when $p = 0.002(\langle k \rangle = 5.84)$.

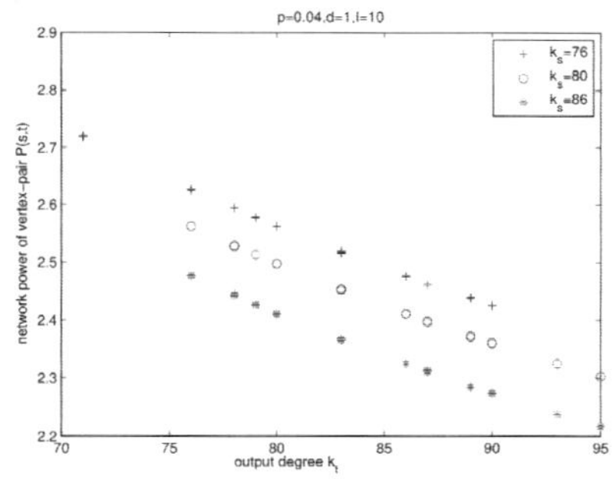

Figure 4. The change of network power of vertex-pair $P(s,t)$ with the change of output vertex degree while the input vertex degree remains unchanged when $p = 0.04(\langle k \rangle = 80.53)$.

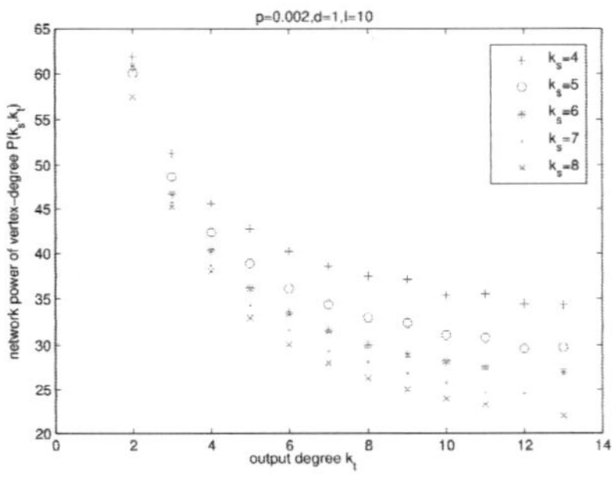

Figure 5. The change of network power of vertex-degree $P(k_s, k_t)$ with the change of output vertex degree while the input vertex degree remains unchanged when $p = 0.002(\langle k \rangle = 5.84)$.

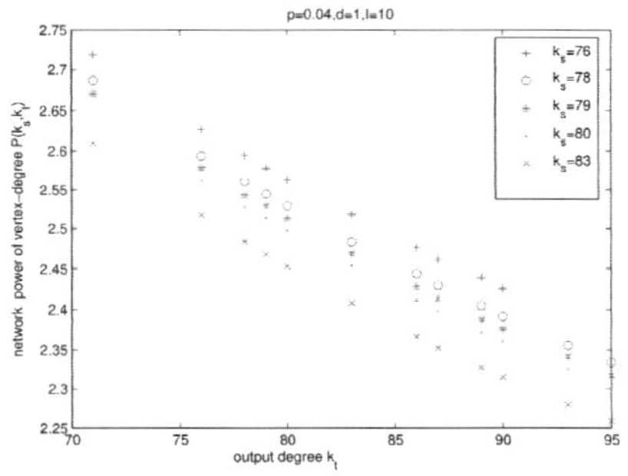

Figure 6. The change of network power of vertex-degree $P(k_s, k_t)$ with the change of output vertex degree while the input vertex degree remains unchanged when $p = 0.04(\langle k \rangle = 80.53)$.

We can find from Figure 1-4 that when the shortest distance between input and output vertices is constant, network power of vertex-pair $P(s,t)$ shows a trend of decreasing along with the increase of the input/output vertex degree k_s/k_t when the output/input vertex degree k_s/k_s remains unchanged. These results can also be observed from Figure 5 and Figure 6, where the network power of vertex-degree $P(k_s, k_t)$ shows a trend of decreasing along with the increasing of the input/output degree k_s/k_t. The reason for the above results is that the number of parallel branches between the input and output vertices is determined by the input and output vertex degree. The greater the k_s and k_t, the more the parallel branches, the smaller $R(s,t)$, the smaller $P(s,t)$ and $P(k_s, k_t)$, such that the current streaming capability of the network will be stronger, whereas weaker.

3.2. The relationship between network power and edge connection probability

In Figure 1 and Figure 3, when the input and output vertex degree k_s, k_t keeps the same, the network power of vertex-pair $P(s,t)$ corresponded to them does not completely identical, but changes in an interval. However, the diversity in Figure 2 and Figure 4 is not obvious, where the network power of vertex-pair $P(s,t)$ is approximately equal. What's more, Figure 5 and Figure 6 show that the change of the network power of vertex-degree $P(k_s,k_t)$ when the connection probability $p = 0.002$ is a curve while the change of the network power of vertex-degree $P(k_s,k_t)$ when the connection probability $p = 0.04$ is almost a straight line. And the network power when the connection probability $p = 0.002$ is obviously larger than the network power when the connection probability $p = 0.04$. These results can be illustrated by network resistance. Assuming that:

- The number of branches between vertex s and vertex t is n. In general, the larger the degree k_s and k_t, the bigger the n.
- The resistance of each branch from vertex s to vertex t is simplified into one same-value R_s(this is true when the size of the complex resistor network is large), then the total resistance R can be formulated as follows:

$$\frac{1}{R} = \underbrace{\frac{1}{R_s} + \frac{1}{R_s} + \cdots + \frac{1}{R_s}}_{n} \tag{5}$$

Thus,

$$R = \frac{R_s}{n} \tag{6}$$

Here R can be $R(s,t)$ or $R(k_s,k_t)$. The smaller n, the more R changes. Such as when $n = 1,2$, $R = R_s, 0.5R_s$, the $R-n$ figures are curves of inversely proportional function. But the larger n and the less R changes, such as when $n = 700,800$, $R = 0.0014R_s, 0.00125R_s$, $R-n$ curves are approximately straight lines. When $p = 0$, the network resistance is the largest and the current streaming capability is the weakest. And when the connection probability p is small, the branch number n from s to t with the same degree (k_s,k_t) fluctuates strongly, so does $P(s,t)$, as shown in Figure 1 and Figure 3: the same input-output degree (k_s,k_t) has different network power $P(s,t)$. As the edge connection probability p keeps on increasing, the average degree of the network $\langle k\rangle$ gradually increases, then n also increases, so $R(k_s,k_t)$ and $P(k_s,k_t)$ reduce. With the increase of n, the fluctuation of n tends to be gentle, thus the change of the network resistance is less and less obvious, and the difference of $P(s,t)$ is smaller and smaller, as shown in Figure 2 and Figure 4: different $P(s,t)$ corresponded to the same (k_s,k_t) gradually concentrate into points. When $p = 1$, all the vertex degrees reach the maximum same value so does n, and $P(s,t)$ and $P(k_s,k_t)$ reach the minimum, $P(k_s,k_t) = P(s,t)$, the current streaming capability is the strongest.

3.3. The relationship between network power and the shortest distance

In this part, we choose proper connection probability p, so that the generated resistor network model has wide range of shortest distance. And the the vertices whose degree are close to the average degree $\langle k\rangle$ are selected as input-output vertices in order to get powerful results. Detailedly, we choose the NW complex small world resistor network with connection probabilities $p = 0.001$, $p = 0.002$ and $p = 0.005$, the corresponding average degrees are $\langle k\rangle = 4.02$, $\langle k\rangle = 5.84$ and $\langle k\rangle = 12.07$, the average shortest distances are $\langle d\rangle = 5.3023$, $\langle d\rangle = 4.1497$ and $\langle d\rangle = 3.0327$, and the shortest distances range among vertices are $d = [1,10]$, $d = [1,7]$ and $d = [1,5]$. Figure 7 to Figure 10 give out the simulation results of network power vs. shortest distance with $k_s = k_t = 4$ when $p = 0.001$, $k_s = k_t = 6$ when $p = 0.002$ and $k_s = k_t = 12$ when $p = 0.005$.

From the three groups of scatters in Figrue 7, Figure 8 and Figure 9, it can be seen that the maximum and minimum value of the network power of vertex-pair $P(s,t)$ have no obvious trends vs. the shortest distance when $d > 1$. Figure 10 shows that when the connection probability p is small, the network power of vertex-degree $P(k_s, k_t)$ increases with the increase of the shortest distance d. However, with the increase of the connection probability p, the difference of the network power of vertex-degree $P(k_s, k_t)$ corresponded to different d are getting smaller, the curves of $P(k_s, k_t)$ vs. d will be an approximately straight line. What's more, $P(k_s, k_t)$ exhibits bigger change when the difference between the value of d and $\langle d \rangle$ is large than the case when d is closes to $\langle d \rangle$. $P(s, t)$ and $P(k_s, k_t)$ with $d = 1$ are both smaller than other cases of d.

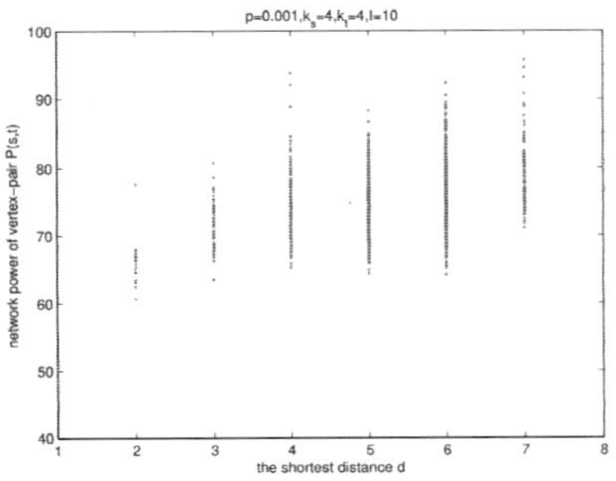

Figure 7. Network power of vertex-pair vs. shortest distance with $k_s = k_t = 4$ when $p = 0.001$.

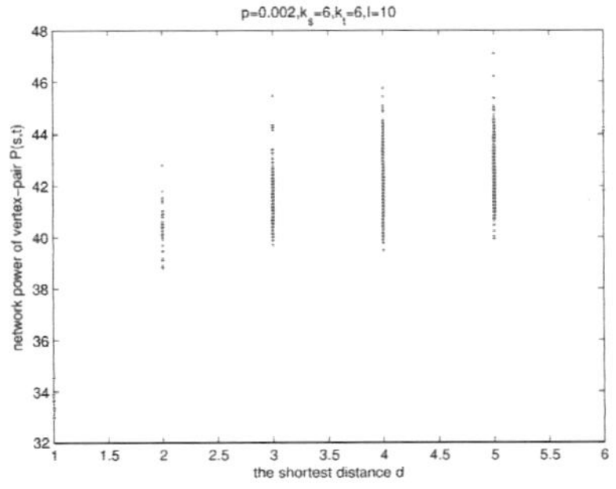

Figure 8. Network power of vertex-pair vs. shortest distance with $k_s = k_t = 6$ when $p = 0.002$.

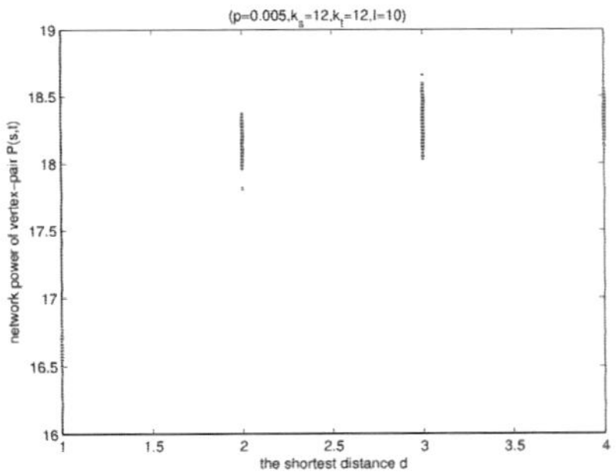

Figure 9. Network power of vertex-pair vs. shortest distance d with $k_s = k_t = 12$ when $p = 0.005$.

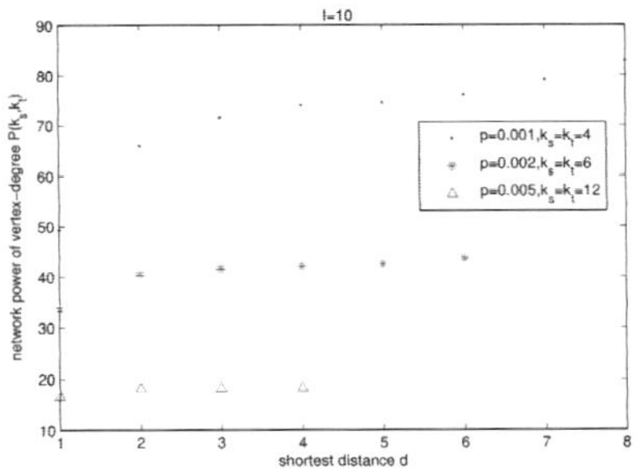

Figure 10. Network power of vertex-degree vs. shortest distance d with different connection probabilities.

For a small resistor network, due to limited number of branches, the total resistance is mainly determined by the branch with the smallest resistance. In a complex network of resistors, however, because of the large number of branches, the total network resistance will not only related to the branch with the smallest resistance but also the number of these branches. Assuming that the shortest distance from vertex s to vertex t is d, the maximum length of loop-free path from s to t is d_{max}, and the simplified resistance between adjacent nodes are all R_0, the property of complex resistor network can be studied by the following simplification:

- Regarding each loop-free path from vertex s to vertex t as a branch.
- The number of the loop-free paths from s to t with length d_i is m_i. The corresponding branch resistance is $d_i R_0$, $d_i \in (d, d_{max})$.

Assuming that $d_i = d + i - 1$, then $i \in (1, d_{max} - d + 1)$. When k_s and k_t are fixed, $m_i = m_i(s,t,d)$. The network resistance of vertex-pair with the same shortest distance $R(s,t,d)$ can be expressed as:

$$R(s,t,d) = \frac{1}{\sum_{i=1}^{d_{max}-d+1} \frac{m_i(s,t,d)}{d_i R_0}} = \frac{R_0}{\sum_{i=1}^{d_{max}-d+1} \frac{m_i(s,t,d)}{d+i-1}} \tag{7}$$

Here we introduce the branching factor:

$$f(s,t,d) = \sum_{i=1}^{d_{max}-d+1} \frac{m_i(s,t,d)}{d+i-1} \tag{8}$$

The network resistance of vertex-degree with the same shortest distance $R(k_s, k_t, d)$ can be written as:

$$R(k_s, k_t, d) = \frac{1}{N(d)} \sum_{(s,t)\in V(k_s,k_t,d)} R(s,t,d) \tag{9}$$

or

$$R(k_s, k_t, d) = \frac{1}{N(d)} \sum_{(s,t)\in V(k_s,k_t,d)} \frac{R_0}{f(s,t,d)} \tag{10}$$

$N(d)$ is number of vertex-pair with the same shortest distance d and vertex degree (k_s, k_t), $V(k_s, k_t, d)$ is the collection of these vertices. For different vertex-pairs with the same degree (k_s, k_t) and shortest distance d, $m_i(s,t,d)$ and d_{max} are usually different, this results in the fluctuation of $P(s,t)$ in Figure 7, Figure 8 and Figure 9. For most of the vertex-pairs in complex resistor network, when d_i is close to the average shortest distance $\langle d \rangle$ of the network, $m_i(s,t,d)$ is large, and when the difference between d_i and $\langle d \rangle$ is large, $m_i(s,t,d)$ is small: $m_i(s,t,d)$ firstly increases and then reduces with the increase of d_i. Thus, for different vertex-pairs with the same vertex degree, when $d = 1$, the value range of $(d_{max} - d + 1)$ for d_i is the maximum, the number of parallel branches is the maximum, f is maximum(Equation 8), $R(k_s, k_t, d)$ of vertex-degree is minimum(Equation 10), the network power of vertex-degree $P(k_s, k_t)$ is minimum, the current streaming capability is the strongest. With the increase of d, the value range of $(d_{max} - d + 1)$ gradually reduces, the number of parallel branches gradually reduces, f reduces, so that the network resistance of vertex-degree $R(k_s, k_t)$ and the network power of vertex-degree $P(k_s, k_t)$ both increase. When d is close to the average shortest distance $\langle d \rangle$, the number of parallel branches decreases, $f(d)$ changes gently, so that $R(k_s, k_t)$ and $P(k_s, k_t)$ change gently, as shown in Figure 10. When $d = d_{max}$,the range of $(d_{max} - d + 1)$ is the smallest, the number of parallel branches reaches the minimum, f is minimum, $R(k_s, k_t)$ and $P(k_s, k_t)$ reach the maximum, the streaming capability is the weakest. Generally speaking, it is the reduction of the number of branches when the shortest distance from the input vertex to the output vertex increases, which leads to the the increase of network resistance and network power.

4. Summary and Outlook

In this paper, we established an undirected complex resistor network based on NW small world network model. The relationship between network current streaming capability and network structure is studied. By simulation and analytical analysis, we got the following results:

a) Complex network resistance and power have a decreasing trend with the increase of the input or output vertex degree.

b) With the increase of edge connection probability, network resistance and power decrease, and the current streaming capability is enhanced. At the same time, the differences between the network resistance and power of vertex-pair and vertex-degree with different connection probabilities are getting smaller.

c)With the increase of the shortest distance, the network resistance and power both increase, and the current streaming capability is weakened.

The underlying reason for the above results is also studied. And in the analysis of network resistance and power with different shortest distance, we gave a simplified formula for network resistance, and introduced a branching factor which could be a new statistical parameters in the analysis of complex resistor network.

More research on the local performance of different complex resistance network will be done in our future work, such as the study of the edge-current, edge-power, vertex-current, vertex-power, the specific form of $m_i = m_i(s,t,d)$ in branch factor $f = f(s,t,d)$. These local study of complex resistor network can provide useful estimation of the importance of different vertex and edge.

Acknowledgement

This work is supported by the National Science Foundation of China under grant Nos. 11147155, 61174216, 61074091 and 11303020, and Education Commission of Hubei Province of China under grant No. Q20121311, Natural Science Foundation of CTGU under grant No. KJ2008B033.

References

[1] Wasserman S, Faust K 1994 *Social Network Analysis* (Cambridge University Press, Cambridge)
[2] Albert R, Barabasi A L 2002 *Rev. Mod. Phys.* 74 47
[3] Newman M E J 2003 *SIAM Review* 45 167
[4] Newman M E J 2010 *Network: An Introduction* (Oxford University Press, New York)
[5] Klein D J, Randicc M 1993 Resistance Distance *Journal of Mathematical Chemistry* 12(1): 81-95
[6] Flake G W, Lawrence S, Giles C L, et al. 2002 Self-organization and identification of veb communities *IEEE computer* 35(3):66-70
[7] Wilkinson D M and Huberman B A 2004 A method for finding communities of related genes *PNAS* 101 5241
[8] Newman M E J and Girvan M 2004 Finding and evaluating community structure in networks *Phys. Rev. E* 69 026113
[9] Danon L, Duch J, Diaz-Guilera A and Arenas A 2005 Comparing Community Structure Identification *J. Stat. Mech.* P09008
[10] Newman M E J 2013 Community detection and graph partitioning *Europhysics Letter* 103 28003
[11] Newman M E J 2013 Spectral methods for community detection and graph partitioning *Phys. Rev. E* 88 042822
[12] Schuetz P and Caflisch A 2008 Efficient modularity optimization by multistep greedy algorithm and vertex mover refinement *Phys. Rev. E* 77 046112
[13] Chang Y F and Ping Y 2011 Emerging cluster analysis of SCI journals and its efficiency *Int. J. Mod. Phys. C* 22(10):1247-1256
[14] Wu F, Huberman B A 2004 Finding communities in linear time: A physics approach *Eur. Phys. J. B.* 38 331-338
[15] Wu F Y 2004 Theory of resistor networks: the two-point resistance *J. Phy. A:Math. Gen* 37 6653
[16] Izmailian N Sh and Ming-Chang Huang 2010 Asymptotic expansion for the resistance between two maximally separated nodes on an M by N resistor network *Phys. Rev. E* 82 011125
[17] Izmailian N Sh, Kenna R and Wu F Y 2014 The two-point resistance of a resistor network: a new formulation and application to the cobweb network *J. Phy. A:Math. Gen* 47 035003
[18] Zhi-zhong Tan, Ling Zhou and Jian-hua Yang 2013 The equivalent resistance of a $3 \times n$ cobweb network and its conjecture of an m by n cobweb network *J. Phy. A:Math. Gen* 46 195202
[19] Tan Z Z, Essam J W and Wu F Y 2014 Two-point resistance of a resistor network embedded on a globe *Phys. Rev. E* 90 012130
[20] Newman M E J and Watts D J 1999 Renormalization group analysis of the small-world network model *Phys. Lett. A* 263: 341-346

Measuring microscopic evolution processes of complex networks based on empirical data

Liping Chi[1,2]

[1]Complexity Science Research Center, College of Physical Science and Technology, Central China Normal University, Wuhan 430079, China
[2]Center for Complex Network Research, Department of Physics, Northeastern University, Boston, MA 02115, USA

E-mail: chilp@mail.ccnu.edu.cn

Abstract. Aiming at understanding the microscopic mechanism of complex systems in real world, we perform the measurement that characterizes the evolution properties on two empirical data sets. In the Autonomous Systems Internet data, the network size keeps growing although the system suffers a high rate of node deletion ($r = 0.4$) and link deletion ($q = 0.81$). However, the average degree keeps almost unchanged during the whole time range. At each time step the external links attached to a new node are about $c = 1.1$ and the internal links added between existing nodes are approximately $m = 8$. For the Scientific Collaboration data, it is a cumulated result of all the authors from 1893 up to the considered year. There is no deletion of nodes and links, $r = q = 0$. The external and internal links at each time step are $c = 1.04$ and $m = 0$, correspondingly. The exponents of degree distribution $p(k) \sim k^{-\gamma}$ of these two empirical datasets γ_{data} are in good agreement with that obtained theoretically γ_{theory}. The results indicate that these evolution quantities may provide an insight into capturing the microscopic dynamical processes that govern the network topology.

1. Introduction

There has been an increasing interest in characterizing the evolution properties of complex systems under the framework of network science. Models have been introduced to capture the basic evolution mechanism determining the system structure. The early model of *growth* and *preferential attachment* proposed by Barabasi and Albert in 1999 [1] successful explained the heterogeneous degree distribution in real-world complex systems. Many other models have been built to obtain different dynamical properties observed in real networks, such as fitness [2], aging [3], copy model [4], temporal pattern [5, 6], to name a few. Recent research puts more emphasis on the time effect of complex systems with time-stamped connectivity, such as bursty in email data [7], time aggregation in mobile phone data [8]. The other direction along this line is to perform the measurements directly on evolving systems to elucidate the formation of network structure. Newman measured the clustering coefficient and preferential attachment of scientific collaboration networks in physics and biology and provided a sublinear power-law degree distribution [9, 10]. Jeong et al inspected the role of added external and internal links during the time evolution of real networks [11]. Although similar in spirit of the latter, our approach emphasizes the measurement of evolution properties at a microscopic scale.

In Ref. [12], the model of four elementary processes in network evolution was analyzed theoretically to understand the individual roles of adding and deleting nodes or links, as well as the random and preferential attachment in link creation. In this paper we will employ the model to investigate two empirical data sets: the Autonomous Systems Internet data and Scientific Collaboration data. In addition to understanding their topology, the study is intended to uncover the microscopic dynamical properties of complex systems in real world and provide a way of probing the evolution processes that determine the network structure.

The rest of the paper is organized as follows. Section II briefly introduces the network evolution model in Ref. [12] and the empirical data sets we analyze. In Section III we present the methodology and corresponding results, especially the comparison of degree distribution between empirical data and theory. Conclusion and discussion are offered in the final section.

2. Model and data description

This paper attempts to characterize the evolution mechanism driving the network structure. The network evolution model will be performed on two empirical data sets to estimate the microscopic dynamical processes.

2.1. Model

In Ref. [12], to understand the individual roles of adding and deleting processes in forming network topology, four elementary processes are included at each time step:

(i) Add a new node along with c external links. Each link is attached to the existing node with probability $\pi(k) = a + bk$;

(ii) Add m internal links among existing nodes. The probability of choosing one end of the internal link is given by $f(k) = s + tk$;

(iii) Delete a node with probability r;

(iv) Delete an internal link with probability q.

For simplicity, we assume $s = a$ and $t = b = 1$, which leads to $\pi(k) = f(k) = a + k$. The parameter a is called the random attachment, or initial attractiveness by Dorogovtsev et al [13]. At time t, the number of nodes in the network is $n_t = (1-r)t$ and the number of edges is $m_t = [c - r\langle k\rangle + m(1-q)]t$. According to $\langle k\rangle = \frac{2m_t}{n_t}$, we obtain the average degree

$$\langle k\rangle = \frac{2}{1+r}[c + m(1-q)]. \tag{1}$$

Degree distribution is an important metric to describe the network properties. We assume the degree distribution has the form

$$p(k) \sim k^{-\gamma}\Omega^k, \tag{2}$$

in which Ω is a controllable parameter. If $\Omega \leq 1$, the degree distribution is a power-law distribution with an exponential cutoff. The exponent γ is given by [12]

$$\gamma = 2 + \frac{(1+r)[c(c+m(1-q)) + a(c+m)(1+r)]}{c^2(1-r) + c[m(1-q)(3-r) - ar(1+r)] + m[2m(1-q)^2 - a(1+r)(q+r)]}. \tag{3}$$

When $2 < \gamma < 3$, the degree distribution is typically a heterogeneous distribution. As γ gets much larger, the degree distribution decays rapidly, indicating that the hubs are few and the distribution tends toward homogeneous.

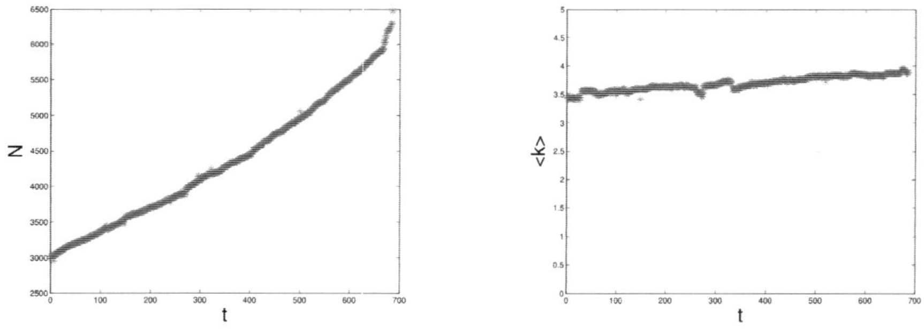

Figure 1. The number of nodes N and the average degree $\langle k \rangle$ on each day for ASN.

2.2. Data description

Two empirical data sets are chosen for analysis.

Autonomous System (AS) Internet data: The data contains 685 daily instances which span an interval from November 8, 1997 to January 2, 2000 [14].

Scientific Collaboration data: The scientific collaboration data consists of all the papers published in APS journals with DOI prefix 10.1103, providing a range of 117 years from July 1893 to Dec 2009 [15, 16]. The data in the considered year is a cumulated result of all the papers and authors from 1893 up to the given year.

3. Method and Results

3.1. Autonomous System Internet data

In the Autonomous System network (ASN), each node is an Autonomous System number and the link represents the communication between Autonomous Systems. Figure 1 presents the number of nodes N and the average degree $\langle k \rangle$ as a function of time. In spite of frequently adding or deleting nodes and links, the network size N keeps growing. However, it is interesting to notice that the average degree $\langle k \rangle$ is almost unchanged.

We plot the degree distribution in Fig. 2 for ASN. The method of logarithmic binning is used in Fig. 2(b) to reduce the noise in the tail. We see that $p(k)$ approximately fits to a power-law degree distribution $p(k) \sim k^{-\gamma}$ with exponent $\gamma_{data} = 2.21$. The subscript *data* denotes the exponent is from the empirical data.

Next we will employ the model to measure the following quantities that characterize the microscopic dynamical processes in the data. First, we will discuss the quantities that describe the topological changes at each time step, which include external links attached to a new node c, added internal links m, probability of node deletion r and probability of link deletion q. These parameters are crucial to understand the evolution processes. Figure 3 reports the time evolution of these parameters. The peak values (ie, most probable numbers) are chosen to approximate the values of these parameters. For ASN, $c = 1.1$, $m = 8$, $r = 0.4$ and $q = 0.81$, as shown in Table 1.

Second, we will focus on the measurement of random attachment a in $\pi(k) = f(k) = a + k$. As stated in Ref. [11], the random attachment is very small and available statistics is insufficient for us to measure it. Hence in this paper we take the random attachment as $a = 0$. It implies that the attachment $\pi(k)$ linearly depends on degree k.

Substituting these quantities into Eq. (3), we obtain the theoretical exponent of degree distribution $\gamma_{theory} = 2.43$. Compared with $\gamma_{data} = 2.21$, the result demonstrates that these quantities can possibly capture the microscopic evolution mechanism of empirical data.

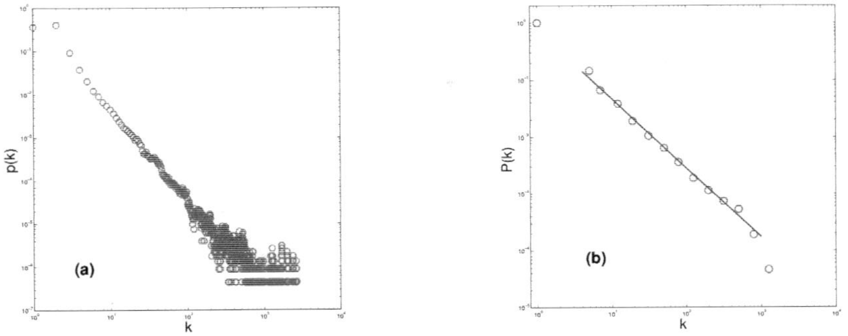

Figure 2. (a) Plot of degree distribution $p(k) \sim k^{-\gamma}$ for ASN with exponent $\gamma_{data} = 2.21$. (b) The method of logarithmic binning is used to reduce the noise in the tail.

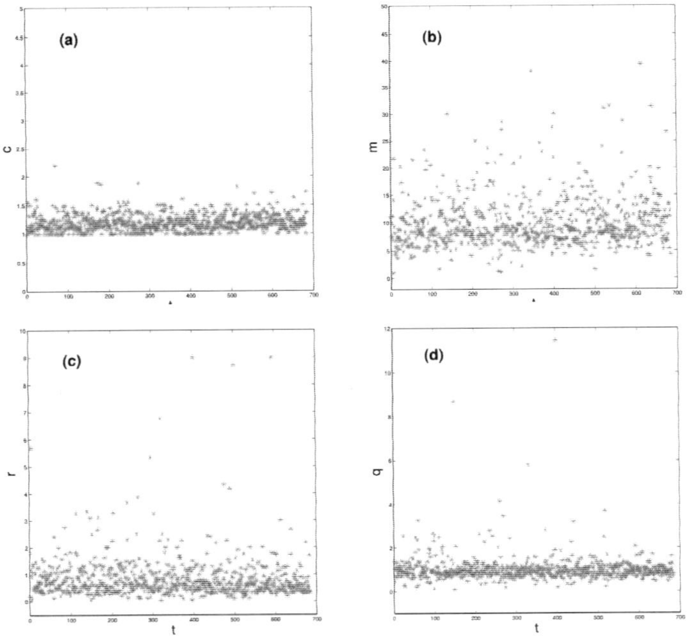

Figure 3. Plot of empirically measured quantities that govern the evolution processes for ASN. (a) added external links attached to a new node c; (b) added internal links m; (c) relative probability of node deletion r and; (d) relative probability of link deletion q. The peak values (ie, most probable numbers) are chosen to approximate the values of these parameters. For ASN, $c = 1.1$, $m = 8$, $r = 0.4$ and $q = 0.81$.

Table 1. Quantities that describing the evolution processes at each time step, including the external links c, internal links m, probability of node deletion r and link deletion q, as well as the power-law exponents obtained theoretically γ_{theory} and empirically γ_{data}.

	c	m	r	q	γ_{theory}	γ_{data}
ASN	1.1	8	0.4	0.81	2.43	2.21
SCN	1.04	0	0	0	3	2.13 - 4.01

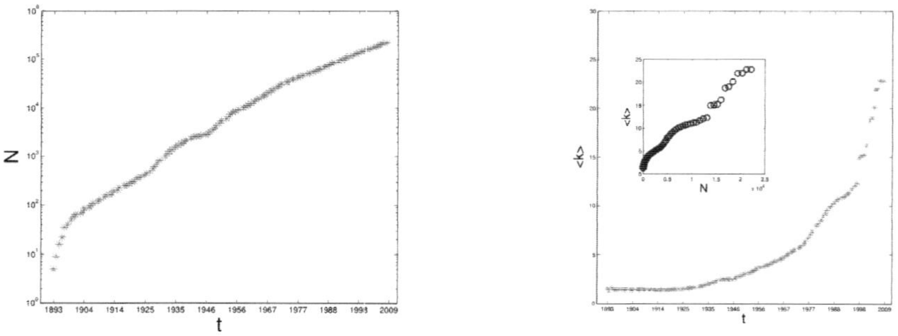

Figure 4. (a) Cumulative number of authors N and (b) the average degree $\langle k \rangle$ in the given year. The inset exhibits that the average degree $\langle k \rangle$ increases linearly with N.

3.2. Scientific Collaboration data

In the Scientific Collaboration network (SCN), the nodes are the authors and the links represent that two authors wrote a paper together. The number of nodes N and the average degree $\langle k \rangle$ as a function of time are plotted in Fig. 4. We see that the network size N grows exponentially with time due to the addition of new authors. The average degree $\langle k \rangle$ is approximately a constant before 1960s and then increases sharply afterwards.

Figure 5 plots the cumulative degree distribution of the given year, assuming the degree distribution is of the form $p(k) \sim (\gamma - 1)k_0^{1-\gamma}(k + k_0)^{-\gamma}$ [18]. The exponents γ_{data} vary for different years, from 4 in 1950 to 2.13 in 2009. The inset exhibits the changes of γ_{data} with time.

Now we calculate the evolution quantities for SCN. It is noticed that the Scientific Collaboration data is a cumulated data. There is no deletion of nodes and links, indicating that $r = 0$ and $q = 0$. The corresponding theoretical exponent of degree distribution is

$$\gamma = 2 + \frac{a + c}{c + 2m}. \tag{4}$$

We plot the external links c and internal links m as a function of time in Fig. 6. The external links c give the average collaborators a new author has. The internal links m represent the number of collaborations between old authors who did not collaborate before. The peak values are $c = 1.04$ and $m = 0$ (see Table 1), which shows that each new author has about one existing old collaborator and there is no collaboration between old authors from the view point of over a century's time.

Providing the random attachment $a = 0$, the theoretical exponent of degree distribution for

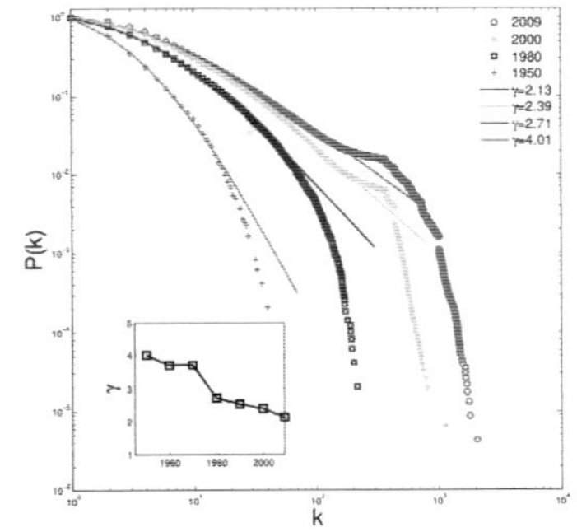

Figure 5. Cumulative degree distribution for SCN, assuming the degree distribution is $p(k) \sim (\gamma - 1)k_0^{1-\gamma}(k + k_0)^{-\gamma}$. The exponents γ_{data} vary for different years, falling into the range of 2.13 to 4.

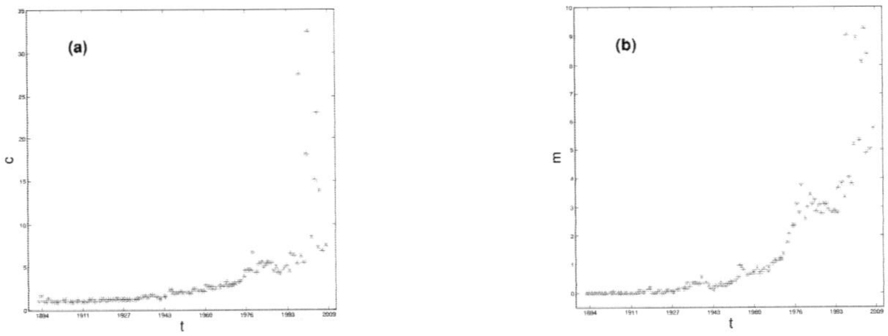

Figure 6. Plot of empirically measured quantities that govern the evolution processes for SCN. (a) added external links attached to a new node c and; (b) added internal links m. The peak values of c and m are 1.04 and 0, respectively.

SCN is $\gamma_{theory} = 3$. The result also exhibits a consistency between the theory and empirical data.

4. Conclusion and discussion

The analysis of microscopic evolution processes of complex networks is performed on two empirical data sets: the Autonomous Systems Internet data and Scientific Collaboration data. A detailed description of the evolution properties has been investigated, including the external links c that a newly added node has, added internal links between existing nodes m, the probability of node deletion r and link deletion q. In the Autonomous Systems Internet data, the network size keeps growing although the system suffers a high rate of node deletion ($r = 0.4$) and link deletion ($q = 0.81$). The average degree $\langle k \rangle$ is approximately a constant as the network size N expands. The external and internal links at each time step for ASN are $c = 1.1$ and $m = 8$. It implies that the internal link plays an important role in network growth. The theoretical and empirical exponents of degree distribution are $\gamma_{theory} = 2.43$ and $\gamma_{data} = 2.21$, showing a good agreement. For the Scientific Collaboration data, it is a cumulated result with $r = q = 0$. The external and internal links are $c = 1.04$ and $m = 0$, correspondingly. A similar consistency in the exponent of degree distribution has been observed for SCN.

The agreement between the empirical data and theory indicates that these quantities can capture the dynamical processes determining the network topology. However, two key questions still remain uncertain. First, in the theoretical analysis of dynamical processes, time t is a continuous variable and being strictly defined. It is unclear how to set time t when analyzing empirical data. In this paper we take one added node as a time step. Second, the time interval is an important parameter in describing the network evolution. If the time interval is too small, there are only several nodes and links in each interval. It is hard to observe the formation of clusters. On the other hand, if the time interval is too large, it contains approximately all the nodes in each interval. It is difficult to inspect the emergence of critical links. It is still an open question on how to choose an optimal time interval to analyze the time-stamped network. In spite of all these challenges, this work provides a way of probing the microscopic evolution mechanism driving the network structure.

Acknowledgement

The author thanks Gourab Ghoshal for valuable discussions and suggestions. This work was partially supported by the Programme of Introducing Talents of Discipline to Universities under Grant No. B08033, and the Fundamental Research Funds for the Central Universities of Central China Norma University (CCNU).

References

[1] Barabási A-L and Albert R 1999 *Science* **286** 509-512
[2] Bianconi G and Barabási A L 2001 *Europhys. Letts.* **54** 436-442
[3] Dorogovtsev S N and Mendes J F F 2000 *Phys. Rev. E* **62** 1842-1845
[4] Krapivsky P L and Redner S 2005 *Phys. Rev. E* **71** 036118
[5] Holme P and Saramäki J 2012 *Physics Reports* **519** 97
[6] Nicosia V, Tang J, Mascolo C, Musolesi M, Russo G and Latora V 2013 *Temporal networks: understanding complex systems*, (Springer Berlin Heidelberg) pp 15-40
[7] Barabási A 2005 *Nature* **435** 207-211
[8] Krings G, Karsai M, Bernhardsson S, Blondel V D and Saramäki 2012 *EPJ Data Science* **1**(4) 1-16
[9] Newman M E J 2003 *Phys. Rev. E* **64** 025102
[10] Newman M E J 2010 *Networks: An Introduction* (Oxford: Oxford University Press)
[11] Jeong H, Neda Z and Barabási A L 2003 *Europhys. Letts.* **61** 567-572
[12] Ghoshal G, Chi L P and Barabási A L 2013 *Scientific Reports* **3** 2920
[13] Dorogovtsev S N, Mendes J F F and Samukhin A 2000 *Phys. Rev. Lett.* **85** 4633
[14] http://snap.stanford.edu/data/as.html
[15] Barabási A L, Jeonga H, Néda Z, Ravasza E, Schubertd A and Vicsekb T 2002 *Physica A* **311** 590-614
[16] Deville P, Wang D, Sinatra R, Song C, Blondel V D and Barabási A L 2014 *Scientific Reports* **4** 4770
[17] Martin T, Ball B, Karrer B and Newman M E J 2013 *Phys. Rev. E* **88** 012814
[18] Ghoshal G and Barabási A L 2011 *Nature Commun.* **2** 394

Measuring fractal dimension of metro systems

S Deng[1], W Li[1,2], J Gu[3], Y Zhu[1,4,5], L Zhao[1] and J Han[1]

[1] College of Physical Science and Technology, Central China Normal University, Wuhan 430079, China.
[2] Max-Planck-Institute for Mathematics in the Science, Inselstr. 22, 04103 Lepzig, Germany
[3] Mathematics and Science College, Shanghai Normal University, Shanghai 200234, China
[4] ISMANS, LUNAM Université, 44 av. Bartholdi, 72000 Le Mans, France
[5] IMMM, UMR CNRS 6283, Université du Maine, 72085 Le Mans, France

E-mail: gitsteven@gmail.com

Abstract. We discuss cluster growing method and box-covering method as well as their connection to fractal geometry. Our measurements show that for small network systems, box-covering method gives a better scaling relation. We then measure both unweighted and weighted metro networks with optimal box-covering method.

1. Introduction

Metro systems are playing an increasingly important role in facilitating urban transportation. In one of our former works by Li *et al* [1], we confronted with the problem of how to measure the fractal dimension of a metro network, we briefly demonstrated how we perform these measurements in the supplementary material of that paper, in this paper we organize our considerations on this issue so as to elucidate some main points that have been studied in the literature on measuring the fractal dimension of a network.

There are two aspects in the structure of a metro network: its geometrical structure and its topological structure. The geometrical structure is a direct map of lines and stations into $\mathbb{R}^2$ or $\mathbb{R}^3$. For such a geometrical object, if there is no a characteristic length scale, fractal geometry is a powerful tool to capture the self-similarity property [2]. An early work studying the geometry fractal property of urban railway system was conducted by Benguigui and Daoud [3], one may be also inclined to do the similar study on metor networks. Geometrical description has its advantages, whereas, it is restricted by practical factors as well as theoretical drawbacks. On the one hand, we found it very difficult to find a complete set of available GPS data; on the other hand, various properties of a metro system can be revealed by constructing its topological network and analysing this network with graph theory. So in the course of studying metro networks, a network accented fractal dimension may always be prefered. The most straightforward approach is to define fractal dimension on the networks.

There are two main definitions of fractal dimension of networks within the current literature: the cluster growing method introduced by Shanker [4], and the box-covering method proposed by Song *et al* [5]. Their studies were based on large size networks, with sizes range from a few thousand to hundreds of thousands. However, metro networks are relatively small, most of which have a size about 100–300 (see figure 2), making scaling relation more difficult to detect. With two definitions of fractal dimension at hand, the metro data enable us to systematically study which is the better way of revealing scaling relation for small size networks, our work may provide a perspective to further studies on small size networks.

The data of 28 metro networks we studied in this paper were manually collected from corresponding official websites (see Appendix). We first observe the characteristics of these metro networks. In section 3 and section 4 we discuss the cluster growing method and the box-covering method with a coherent connection with geometry fractal. Upon these fundamental considerations together with our observations in the measuring processes, we use box-covering method to measure the fractal dimension of major world-wide metro networks.

2. Properties of typical metro networks

Many real-world networks have properties of small-world and scale-free. Small-world means in many actual networks, one only need a very small number of steps to reach one node from any other node. This implies a slow increase of average distance $\langle l \rangle$ of the network with the system size N [6–8]:

$$\langle l \rangle \sim \ln N \quad \text{or} \quad N \sim e^{\langle l \rangle / l_0}, \tag{1}$$

with l_0 a characteristic length. Scale-free property indicates that the degree distribution of a complex network obeys a power-law distribution with an exponent γ in the range $(2, 3)$ [9–11]:

$$P(k) \sim k^{-\gamma}. \tag{2}$$

The degree distributions of metro networks are all very narrow (see figure 1), even an underling scale-free distribution may exist when the system size becomes large, we can't claim scale-free property for such relatively small systems. Figure 2 gives a scattering pattern between metro system size N and average distance $\langle l \rangle$, this pattern is very diverse from what may expect by Eq. (2). Therefore metro networks are *non-small-world* and *non-scale-free*. This is in agree with our daily experience that we don't have a special express train that connects two distant stations directly nor do we have hubs with large degee in a metro system.

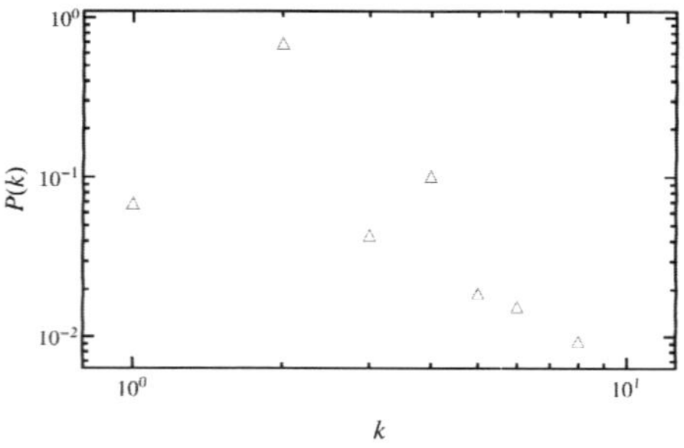

Figure 1. Degree distribution of Paris metro network.

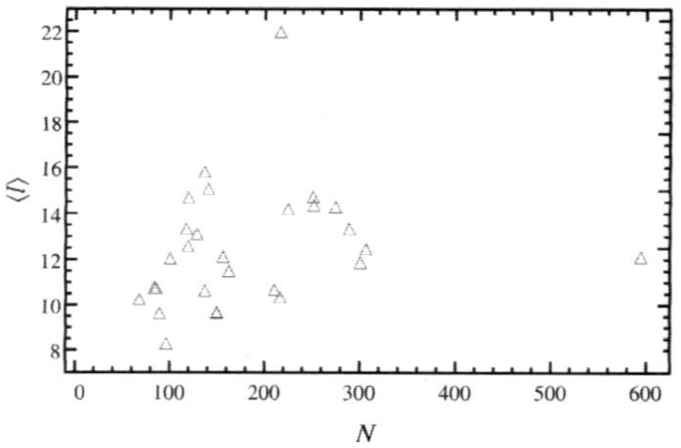

Figure 2. 28 metro networks: system size N versus average distance $\langle l \rangle$.

3. The cluster growing method

For a distance vector $\mathbf{r} = (x_1, x_2, \cdots, x_d)$ in the Euclidean space $\mathbb{E}^d$, the Euclidean distance is defined as the l_2-norm:

$$\|\mathbf{r}\|_2 = \sqrt{x_1^2 + x_2^2 + \cdots + x_d^2}. \tag{3}$$

A fractal object $\mathbf{F}$ embedded in this space has a fractal dimension d_f defined as the scaling exponent of the scaling relation

$$V(r) \sim r^{d_f}, \tag{4}$$

where $V(r)$ is the volum of $\mathbf{F}$ enclosed in the d-sphere of radius r. In discrete space, the equivalence of $\mathbb{E}^d$ is the regular lattices $\mathbb{Z}^d$. A homogeneous complex network characterized by a homogeneous degree distribution can be regarded as a fractal object being embedded inside certain $\mathbb{Z}^d$ [4, 5]. For instance, in a one-dimensional ring consists of N nodes, adding edges in turn with probability p to connect each node N_i with a new node N_{i+m} that is m edges away, then the derived network is an untouched one-dimensional ring if $p = 0$ and a two-dimensional $m \times N/m$ lattice if $p = 1$. For $0 < p < 1$ the resulting network has a dimension between 1 and 2 [4]. In a network, the distance between two nodes is taken as the shortest path length. In this sense, this distance is expressed in $\mathbb{Z}^d$ as the l_1 norm

$$\|\mathbf{n}\| = \|n_1\| + \cdots + \|n_d\|, \tag{5}$$

where $\mathbf{n} = (n_1, n_2, \ldots, n_d)$ is a distance vector in $\mathbb{Z}^d$. The fractal dimension d_f can be defined in a similar manner as Eq. (4) [4]

$$\langle M(r) \rangle \sim r^{d_f}, \tag{6}$$

where $M_i(r)$ is the number of nodes (mass) contained in a disk of radius r centered at node i. $M_i(r)$ generally varies with the seeded center i, so we need to average over many (say $\mathcal{N}$) seed center nodes. For a network consists of a large number of nodes, we can push $\mathcal{N}$ to be a large number and make $\langle M(r) \rangle$ almost independent of the seed nodes we choose. Whereas for our metro networks, the typical number of stations is less than 300 (see figure 2), randomness seems inevitable. We can always average over all nodes to avoid this randomness, but seed nodes taken from the periphery will bring in a large bias and break the scaling relation (6), as shown in figure 3 and Ref. [12]. As will be demonstrated in section 4, the scaling in the form of (6) doesn't even exist for small-world networks. So we will not utilize cluster growing method to measure fractal dimension of metro networks.

4. The box-covering method

In fractal geometry, another method of measuring fractal dimension is the box-covering method, in which we count the number of boxes N_B that is needed to cover a fractal object $\mathcal{F}$, given the scaling relation between N_B and box size s

$$N_B \sim s^{-d_B}, \tag{7}$$

d_B is then the fractal dimension of $\mathcal{F}$ [2]. On a network the similar tiling is illustrated in figure 4. The box size is defined as chemical distance l_B, which means that the distance of any two nodes within this box is less than l_B [5, 13]. Denoting the minimum number of boxes that is needed to cover the entire network by N_B, the fractal dimension d_B is then given by [2, 5]

$$N_B \sim l_B^{-d_B}, \tag{8}$$

provided that this scaling relation exists, and the network is said to be fractal. In fractal geometry, self-similarity refers to the invariance of a fractal object under scaling transformation. On a network, the self-similarity is the invariance of degree distribution $P(k)$ under iterative coarse graining renormalization procedure as defined in Ref. [5]. Topological fractality and self-similarity both manifest in some networks, such as WWW and actor networks [5, 14]. This is not a surprise, but Song *et al* in [5, 14, 15] found that non-fractal Internet network is actually

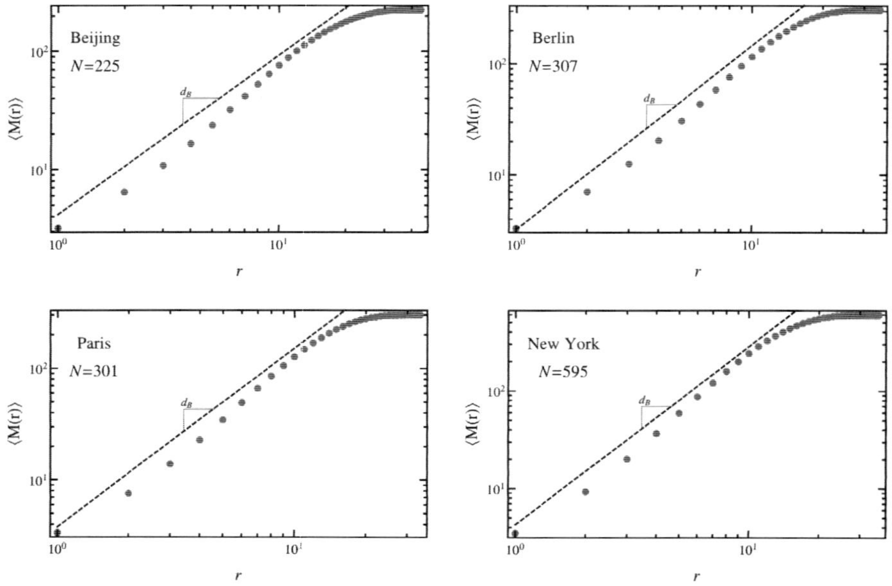

Figure 3. Cluster growing results for metro networks of Beijing, Berlin, Paris and New York: $\langle M(r) \rangle$ versus r, we cut off r at d_{max}, the diameter of the network. The dashed lines are plotted with slops d_B obtained by box-covering method (see table 1).

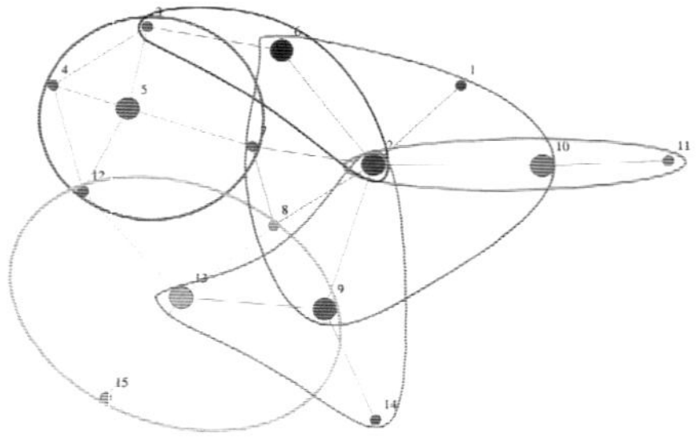

Figure 4. A sample network consists of 15 nodes covered by 6 boxes of size $l_B = 3$.

self-similar. Thus, unlike fractal geometry, fractality and self-similarity are considered to be distinct in complex networks.

The fractality of complex networks is exposed to be related to their underling tree structure called skeleton [16]. A skeleton is a spanning tree consists of edges with largest bewteenness centrality, the remaining edges are called shortcuts. If a network is constructed in a modular way, the edges with the highest betweenness centrality then provide the main communication routes between the modules. In [16] Goh *et al* found that for fractal networks, the number of boxes that is needed to cover the skeleton is almost the same as that of needed by the original fractal network, which implies that the fractality stems from the skeleton tree. In this reasoning,

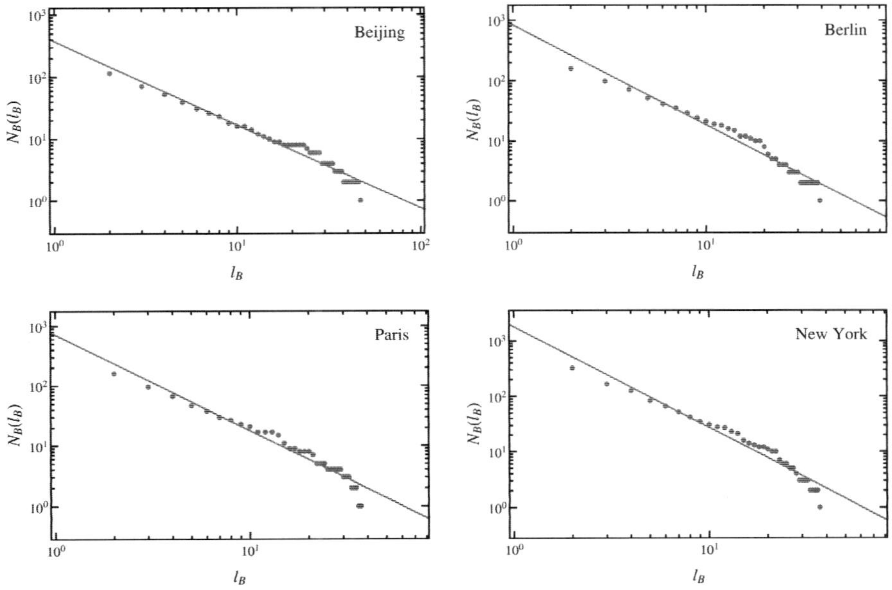

Figure 5. Box-covering results for metro networks of Beijing, Berlin, Paris and New York. The fitted slops d_B are listed in table 1.

a skeleton tree can be considered as being embedded inside certain lattice of $\mathbb{Z}^d$, as discussed in section 3, such a structure is fractal. We can imagine the original network growing as the skeleton tree branches, as the tree branches, some local shortcuts are added which will preserve the "grand picture" of the tree. The dimension we measure on the original network is thus generally the dimension of its skeleton tree. If the locality of wiring the shortcuts is broken, the "grand picture" of the skeleton is lost, and fractality is erased [14, 16].

For a network with small-world or scale-free property that is charecterized by a broad degree distribution, by using Eq. (4), typically, when choosing a random seed as the disk center, it is very probable to include hubs into the disk. Then $M(r)$ will grow in general as $\sim e^{r/l_0}$, which is not a power-law, hence Eq. (6) and (8) are not equivalent. This is due to the over-covering of hub nodes. On the other hand, box-covering method covers all nodes on the same footing, once a node (include hub) is covered, it cannot be covered again [5]. This distinction between the two methods makes cluster growing method bias in unveiling the essential tree structure when working with inhomogeneous networks. However, for a homogeneous network that can be regarded as being embedded inside a $\mathbb{Z}^d$ lattice, two measuring methods are equivalent, $d_f = d_B$ (cf. figure 3, as can be seen, for metro networks, the first few dots obtained by using cluster growing method are very close to the dashed line obtained by box-covering method).

In section 3, we encountered fitting problem when try to measure d_f. Although metro networks are not scale-free or small-world, their relatively small system size still cause fitting difficulties, we hope box-covering method can tame this problem. The measuring results ought to be deterministic due to the nature of this method, however, choosing the optimal minimum number of boxes to cover the network is NP hard, there are 2^N possibilities in the solution space for a system of size N. A former technique to attack this difficulty is deviced by Song *et al* [17], they mapped this problem to graph coloring problem in which the boxes needed are the colors needed to colorize the dual network (see Ref. [17] for details). This greedy coloring

Table 1. Fractal dimension of 28 cities' metro networks

City	d_B	City	d_B	City	d_B
Barcelona	1.469 ± 0.052	London	1.452 ± 0.060	Seoul	1.509 ± 0.025
Beijing	1.347 ± 0.040	Madrid	1.409 ± 0.056	Shanghai	1.342 ± 0.069
Berlin	1.655 ± 0.056	Melbourne	1.231 ± 0.039	Shenzhen	1.185 ± 0.027
Boston	1.250 ± 0.062	Mexico	1.390 ± 0.040	Taipei	1.180 ± 0.056
Busan	1.161 ± 0.031	Milan	1.552 ± 0.062	Tokyo	1.424 ± 0.063
Chicago	1.319 ± 0.071	Montreal	1.090 ± 0.038	Valencia	1.255 ± 0.036
Delhi	1.192 ± 0.045	Moscow	1.462 ± 0.093	Vienna	1.279 ± 0.078
Guangzhou	1.246 ± 0.027	New York	1.824 ± 0.068	Washington	1.243 ± 0.072
Hamburg	1.410 ± 0.068	Osaka	1.445 ± 0.081		
Hong Kong	1.233 ± 0.045	Paris	1.600 ± 0.061		

algorithm is effecient and can give results nearly optimal. Some alternatives include several burning algorithms, where boxes that apparently do not belong to the solution are sequentially excluded [17]. Christian *et al*[1] [13] extended burning algorithm by firstly burning unnecessary boxes and nodes which reduces the solution space considerably, then accompany by spliting the resulting system into subsystems and candidate configurations. In this manner, after several iterations, all candidates are taken into account and the optimal solution can be picked out. But there is a subtlety here, in [13] boxes are calculated with a central node or edge, the box radius r_B and box size l_B is connected by $r_B = (l_B - 1)/2$ for a central node and $r_B = l_B/2$ for a central edge, this may lead to one more box for loops. For example, in a chain of four nodes (1-2-3-4-1), all nodes are seperated from other nodes at most by 2 edges ($l_B = 3$), we only need one $l_B = 3$ box to cover the chain; however, we cannot find a box with radius one ($r_B = (l_B - 1)/2 = 1$) which contains all nodes, and we need one more box to cover all nodes. Excepting for this subtlety, we can take this algorithm to be optimal, and we will use it extensively hereafter. Some measuring examples are shown in figure 5. Unlike in figure 3, here we don't need to worry about where to cut off for our measurement. We list the full measuring results in table 1, dimension values are estimated by linear regression, and the errors are estimated as the fitting error.

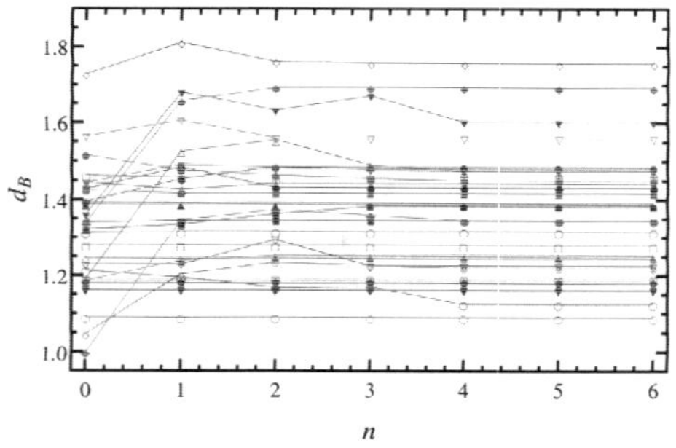

Figure 6. The weighted dimension of 28 metro networks vs. the exponent n.

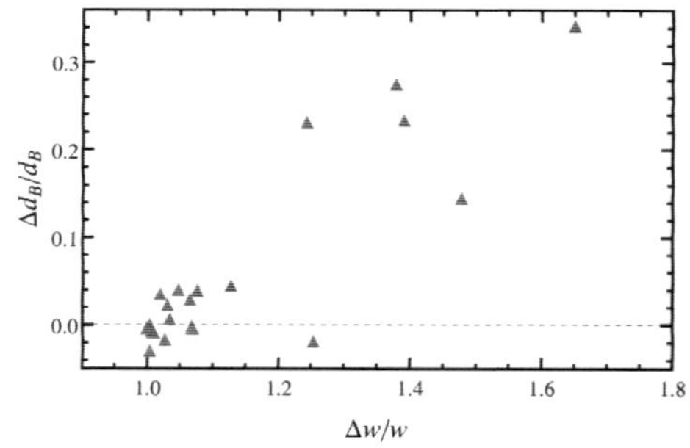

Figure 7. The impact of weight on the change of dimension, here $n = 1$.

[1] We thank for their courtesy for sharing their implementation code.

Many real-world networks are weighted, including many metro networks. In metro networks, the weight of an edge is the number of lines it contains. If there are many such weighted edges, not only the weighted edges can sustain a larger conductance, but also the transition between different lines of the metro will be largely improved, one can then get easier to transport to his/her destination. This empirical experience encourages us to regard edge weights as an evaluation of the shrink of the distance. The traditional interpretaion of edge weight is the distance of this edge: an unweighted edge has distance 1 and an edge of weight w represent a distance of w. We see this is not quite right in metro networks, a more general connection between the distance and weight is given by [18]

$$l_{ij} = \frac{1}{w_{ij}^n}. \tag{9}$$

In order to "shrink" the distance, we can choose $n > 0$, when $n = 0$ we retain the unweighted case. In figure 6-7, we examine the influence of weights on dimension values. Figure 6 shows that shinking the distance according to Eq. (9) will lead to a change of dimension value. As we increase n, most dimension values increase first then fall down, all dimension values tend to a constant if n is large enough. In figure 7, taking $n = 1$, the comparison between change of dimension value versus change of weight indicates a increasing trend between these two changes. Unweighted networks should fall on the dashed line, but it is interesting to see a few dots go below the dashed line even though they correspond to weighted network.

5. Summary

We discussed cluster growing method and box-covering method, and their connection with fractal geometry. According to these discussions, the two methods are equivalent for homogeneous systems whereas are distinct for inhomogeneous systems. The non-small-world and non-scale-free properties of metro networks allow us to measure fractal dimension with cluster growing method. But our measurements show that for such small systems, box-covering method are still more powerful in revealing the scaling relation.

Acknowledgments

This work was partially supported by the Programme of Introducing Talents of Discipline to Universities under Grant No. B08033.

Appendix

Data resources

Our data of 28 world major metros, updated till Nov. 7, 2013, can be retrieved from the following web sites:

- Barcelona: `www.stm.info/en/info/networks/metro`
- Beijing: `www.bjsubway.com/`
- Berlin: `www.bvg.de/index.php/en/index.html`
- Boston: `www.mbta.com/schedules_and_maps/subway/`
- Busan: `www.subway.busan.kr/china/main/`
- Chicago: `www.metrochicago.com/`
- Delhi: `www.delhimetrorail.com/`
- Guangzhou: `www.gzmtr.com/`
- Hamburg: `www.hvv.de/en/`
- Hong Kong: `www.mtr.com.hk/eng/homepage/cust_index.html`
- London: `www.tfl.gov.uk/modalpages/2625.aspx`
- Madrid: `www.metromadrid.es/en/`
- Melbourne: `www.metrotrains.com.au/`

- Mexico: `mexicometro.org/`
- Milan: `www.atm.it/en/Pages/default.aspx`
- Montreal: `www.stm.info/en/info/networks/metro`
- Moscow: `engl.mosmetro.ru/`
- New York: `new.mta.info/`
- Osaka: `www.kotsu.city.osaka.lg.jp/foreign/english/`
- Paris: `www.parismetro.com/`
- Seoul: `www.smrt.co.kr/main/index/index002.jsp`
- Shanghai: `www.shmetro.com/`
- Shenzhen: `www.szmc.net/`
- Taipei: `english.trtc.com.tw/`
- Tokyo: `www.tokyometro.jp/en/`
- Valencia: `www.metrovalencia.es/page.php`
- Vienna: `www.wienerlinien.at/eportal/ep/home.do?tabId=0`
- Washington: `www.wmata.com/`

References

[1] Li W, Gu J, Liu S, Zhu Y, Deng S, Zhao L, Han J and Cai X 2014 Optimal Transport in Worldwide Metro Networks *preprint* arXiv:1403.7844
[2] Mandelbrot B B 1983 *The Fractal Geometry of Nature* vol 173 (New York: Macmillan)
[3] Benguigui L and Daoud M 1991 *Geog. Anal.* **23** 362–368
[4] Shanker O 2007 *Mod. Phys. Lett.* B **21** 321–326
[5] Song C, Havlin S and Makse H A 2005 *Nature* **433** 392–395
[6] Watts D J and Strogatz S H 1998 *Nature* **393** 440–442
[7] Kleinberg J M 2000 *Nature* **406** 845–845
[8] Watts D J 2004 *Six Degrees: The Science of a Connected Age* (New York: W.W. Norton & Company)
[9] Barabási A L and Albert R 1999 *Science* **286** 509–512
[10] Newman M E 2003 *SIAM Rev.* **45** 167–256
[11] Dorogovtsev S N and Mendes J F 2013 *Evolution of Networks: From Biological Nets to the Internet and WWW* (Oxford: Oxford University Press)
[12] Long G and Cai X 2009 *Chin. Phys. Lett.* **26** 088901
[13] Schneider C M, Kesselring T A, Andrade Jr J S and Herrmann H J 2012 *Phys. Rev.* E **86** 016707
[14] Song C, Havlin S and Makse H A 2006 *Nat. Phys.* **2** 275–281
[15] Gallos L K, Song C and Makse H A 2007 *Physica* A **386** 686–691
[16] Goh K I, Salvi G, Kahng B and Kim D 2006 *Phys. Rev. Lett.* **96** 018701
[17] Song C, Gallos L K, Havlin S and Makse H A 2007 *J. Stat. Mech: Theory Exp.* **2007** P03006
[18] Wei D J, Liu Q, Zhang H X, Hu Y, Deng Y and Mahadevan S 2013 *Sci. Rep.* **3**

Analysis on the synchronized network of Hindmarsh–Rose neuronal models

Youceng Feng [1], **Wei Li** [2]
[1] College of Electronics and Information Engineering, South-Central University for Nationalities, Wuhan, 430074, China
[2] Complexity Science Center, and Institute of Particle Physics, Hua-Zhong Normal University, Wuhan, 430079, China
E-mail: fengyc@mail.scuec.edu.cn, liw@mail.ccnu.edu.cn

Abstract. We studied a network of pulse-coupled Hindmarsh-Rose neurons and discovered that all that matters for the onset of complete synchrony is the number of signals, *k*, received by each neuron. This is independent of all other details of the network structure.

1. Introduction

The problem of synchronization of coupled neurons of the system is the key to study the brain information processing. The collective behaviour of various coupled neural networks is the focus of the research [1]. Model studies of neuronal synchronization can be separated into those where threshold models of integrate-and-fire type are used [2] and those where conductance-based spiking and bursting models are employed [3]. In contrast to coupled spiking neurons, whose synchronous dynamics is relatively simple, interacting bursting neurons may exhibit different forms of synchrony [4]. Typically, burst synchronization arises at a low coupling strength whereas complete synchrony, which involves both burst and spike synchronization regimes, requires a stronger coupling. The coupling is pulsatile and often modeled as a static sigmoidal nonlinear input-output function with a threshold and saturation [5].

One important question about interacting neurons with such linear and pulsatile couplings is that of complete synchronization. What are the conditions for the stability of the complete synchronous state, especially with respect to coupling strengths and coupling configurations of the network? This problem was intensively studied for linearly coupled networks of bursting neurons [6, 7], and more generally, of limit-cycle and chaotic oscillators [8, 9]. In particular, it has been shown that synchrony in such networks strongly depends on the structure and size of the network.

The purpose of this letter is to investigate the synchronous stability of a network of pulse-coupled Hindmarsh-Rose (HR) neurons [10]. We discovered that all that matters for the onset of complete synchrony is the number of signals, *k*, received by each neuron.

2. The Hindmarsh-Rose neuronal model

The HR model of neuronal activity is aimed to study the spiking-bursting behaviour of the potential membrane experimentally observed in a single neuron. In fact, the original Hodgkin-Huxley equations cannot describe bursting but only reproduce spiking, while the HR model describes the bursting by taking into account slow currents. Concerning HR model, the relevant variable is the membrane potential, $x(t)$ (dimensionless units). There are two more variables, $y(t)$ and $z(t)$, which take into account the transport of ions across the membrane through the ion channels. The transport of sodium and potassium ions occurs through fast ion channels, and its rate is described by $y(t)$, labelled as the spiking variable. The transport of other ions occurs by slow channels, and is taken into account through $z(t)$, the bursting variable. The corresponding mathematical model is a system of three first-order ordinary differential equations with dimensionless dynamical variables $x(t)$, $y(t)$ and $z(t)$. They read:

$$\dot{x} = ax^2 - x^3 - y - z, \quad \dot{y} = (a+\alpha)x^2 - y, \quad \dot{z} = \mu(bx + c - z) \cdot \tag{1}$$

It is well know that the HR neuronal model presents a chaotic behaviour for many neuronal parameters. For the variables in Equation (1), the physical parameters generating a chaotic behaviour are: a=2.8, α=1.6, c=5, b=9, μ=0.001. Figure 1 shows the corresponding time series. The chaotic transition between bursting periodic and spiking periodic solutions, the strange attractor and the time evolution of the action potential, can be found in Ref. [11].

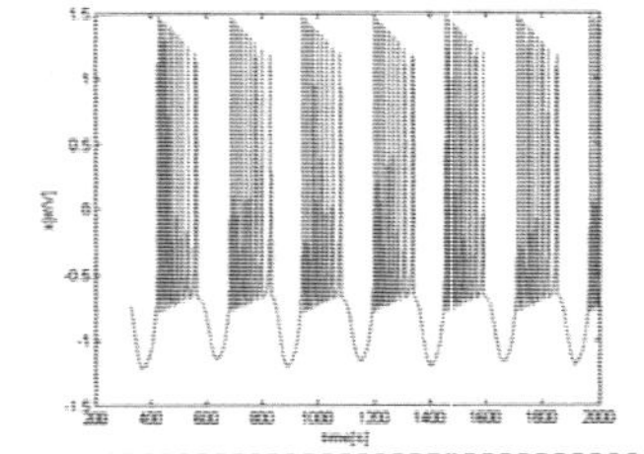

Figure 1. The square-wave bursting of the HR model and the corresponding time series.

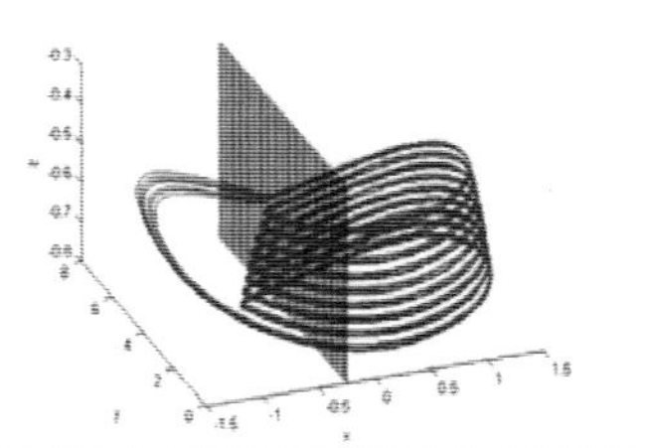

Figure 2. The phase graph of the HR model and the synaptic threshold Θ_s.

3. The network of identical Hindmarsh-Rose neuronal models

Consider now a network of n synaptically coupled HR models. The equations of motion are the following:

$$\dot{x}_i = ax_i^2 - x_i^3 - y_i - z_i - g_s(x_i - V_s)\sum_{j=1}^{n} c_{ij}\Gamma(x_j),$$

$$\dot{y}_i = (a+\alpha)x_i^2 - y_i,$$

$$\dot{z}_i = \mu(bx_i + c - z_i) \cdot \tag{2}$$

Here, $i, j = 1,\ldots,n$, the neurons are identical and the synapses are fast and instaneous. The parameter g_s is the synaptic coupling strength. The reversal potential $V_s > x_i(t)$ for $\forall\, x_i$ and $\forall\, t$, i.e., the synapse is excitatory. The synaptic coupling function is modeled by the sigmoid function $\Gamma(x_j) = 1/\left[1+\exp\{-\lambda(x_j - \Theta_s)\}\right]$ (a limiting version of $\Gamma(x_j)$ is the Heaviside function). This oft-used coupling form was called fast threshold modulation by Somers and Kopell [12]. The threshold Θ_s is chosen such that every spike in the single neuron burst can reach a threshold (see Figure 2).

Hereafter, $\Theta_s = -0.25$ and $V_s = 2$. $\mathbf{C} = (c_{ij})$ is the $n\times n$ connectivity matrix: $c_{ij} = 1$ if neuron i is connected to neuron j, $c_{ij} = 0$ otherwise, and $c_{ii} = 0$. Matrix **C** can be asymmetric *if* both mutual and unidirectional couplings are allowed.

4. Transversal stability of the synchronization manifold

We require equal row sums $k=\sum_{j=1}^{n} c_{ij}, i=1,\ldots,n$. This requirement is a necessary condition for the existence of the synchronous solution, namely, the invariance of hyperplane $D=\{\xi_1(t)=\xi_2(t)=\cdots=\xi_n(t)\}$, $\xi_i=(x_i,y_i,z_i)$ and $i=1,\ldots,n$.

In fact, the equal row-sum property implies a network where each cell has the same number *k* of inputs from other neurons. Synchronous behaviour on the manifold *D* is generated by the system:

$$\begin{aligned}\dot{x}&=ax^2-x^3-y-z-kg_s(x-V_s)\Gamma(x),\\ \dot{y}&=(a+\alpha)x^2-y,\\ \dot{z}&=\mu(bx+c-z)\,.\end{aligned} \tag{3}$$

By introducing the difference between the neural oscillator coordinates $\xi_{ij}=x_j-x_i$, $\eta_{ij}=y_j-y_i$, $\zeta_{ij}=z_j-z_i$, $i,j=1,\ldots,n$, and with the limit of infinitesimal differences, we can change the stability equations for the transverse perturbations to the synchronization manifold *D*:

$$\begin{aligned}\dot{\xi}_{ij}&=(2ax-3x^2)\xi_{ij}-\eta_{ij}-\zeta_{ij}-kg_s\Gamma(x)\xi_{ij}\\ &\quad+g_s(V_s-x)\Gamma_x'(x)\left(k\xi_{ij}+\sum_{h=1}^{n}\{c_{jh}\xi_{jh}-c_{ih}\xi_{ih}\}\right),\\ \dot{\eta}_{ij}&=2(a+\alpha)x\xi_{ij}-\eta_{ij},\\ \dot{\zeta}_{ij}&=\mu(b\xi_{ij}-\zeta_{ij})\,.\end{aligned} \tag{4}$$

The derivatives are calculated at the point $\xi=0$, $\eta=0$, $\zeta=0$, and $\{x(t),y(t),z(t)\}$ corresponds to the synchronous bursting solution defined via system (3).

The first coupling term $S_1=kg_s\Gamma(x)\xi_{ij}$ accounts for the number of inputs *k*. At the same time, the contribution of the second coupling term $S_2=g_s(V_s-x)\Gamma_x'(x)(\cdot)$ depends on the coupling configuration. Note that the term $\sum_{h=1}^{n}\{c_{jh}\xi_{jh}-c_{ih}\xi_{ih}\}$ is the same as for linear coupling [8].

In terms of the original variables x_i, the corresponding coupling matrix $\mathbf{G}=\mathbf{C}-k\mathbf{I}$ is the Laplacian of the connected graph, except for a sign change.

It is well known that $\mathbf{G}$ has one zero eigenvalue γ_1 and all other eigenvalues have nonpositive real parts. If the coupling is mutual, $\mathbf{G}$ is symmetric and all eigenvalues are real. For simplicity, suppose that the eigenvalue γ_2 with the largest real part is simple. Then, by applying the linear transform that diagonalizes $\mathbf{G}$ to Eq. (4), we obtain the stability equation for the most unstable transverse mode:

$$\begin{aligned}\dot{\xi}&=(2ax-3x^2)\xi-\eta-\zeta-\Omega(x)\xi,\\ \dot{\eta}&=2(a+\alpha)x\xi-\eta,\\ \dot{\zeta}&=\mu(b\xi-\zeta),\end{aligned} \tag{5}$$

where $\Omega(x)=kg_s\Gamma(x)-g_s(V_s-x)\Gamma_x'(x)(k+\gamma_2)$. System (5) is an analogy of the Master Stability function [8] for synaptically coupled networks (2). For a given $x(t)$, $\Gamma(x)$ is determinate. For different network structures, the value of γ_2 is variable but determined. The synchronization threshold of g_s is inversely proportional to the number of incoming signals *k* [13].

5. Conclusion

In this paper, we observe the dynamics of HR model and construct coupling network with identical HR neurons. Through the analysis of transversal stability of the synchronization manifold, we obtain that the synchronization condition depends on the number of inputs k but does not rely on the connectivity matrix. The onset of complete synchrony is independent of all other details of the network structure.

References

[1] Ivanchenko M V et al 2004 *Phys.Rev.Lett.* **93** 134101
[2] Gerstner W and Van Hemmen J L 1993 *Phys.Rev.Lett.* **71** 312
[3] Izhikevich E M 2001 *SIAM Rev* **43** 315
[4] Dhamala M, Jirsa V K and Ding M 2004 *Phys.Rev.Lett.* **92** 028101
[5] Somers D and Kopel N 1993 *Biol.Cybernet* **68** 393
[6] Sherman A 1994 *Bull.Math.Biol.* **56** 811
[7] Dhamala M, Jirsa V K and Ding M 2004 Phys.Rev.Lett. **92** 074104
[8] Pecora L M and Carroll T L 1998 *Phys.Rev.Lett.* **80** 2109
[9] Belykh V N, Belykh I V and Hasler M 2004 *Physica D* **195**
[10] Hindmarsh J L and Rose M 1984 *Proc.R.Soc.B* **221** 87
[11] Dtchetgnia Djeundam S R et al. 2013 *Chaos* **23** 033125
[12] Somers D and Kopell N 1993 *Biol.Cybernet.* **68** 393
[13] Belykh V N, Belykh I V and Halsler M 2004 *Physica D* **195** 159

Recent Progress in Some Active Topics on Complex Networks

J Gu[1,2], Y Zhu[3,2], L Guo[4], J Jiang[5], L Chi[2], W Li[2,*], Q A Wang[3], and X Cai[2]

[1] School of Science, Jiangnan University, Wuxi 214122, China
[2] College of Physical Science and Technology, Huazhong Normal University, Wuhan 430079, China
[3] Laboratoire de Physique Statistique et Systemes Complexes, ISMANS, 44 ave. Bartholdi, Le Mans 72000, France
[4] School of Mathematics and Physics, China University of Geosciences (Wuhan), Wuhan 430074, China
[5] Research Center of Nonlinear Science and College of Mathematics and Computer Science of Wuhan Textile University, Wuhan 430200, China
[*] Corresponding author

E-mail: liw@mail.ccnu.edu.cn

Abstract. Complex networks have been extensively studied across many fields, especially in interdisciplinary areas. It has since long been recognized that topological structures and dynamics are important aspects for capturing the essence of complex networks. The recent years have also witnessed the emergence of several new elements which play important roles in network study. By combining the results of different research orientations in our group, we provide here a review of the recent advances in regards to spectral graph theory, opinion dynamics, interdependent networks, graph energy theory and temporal networks. We hope this will be helpful for the newcomers of those fields to discover new intriguing topics.

1. Introduction

Complex network describes the interaction among the elements of complex systems, which consider particular nature of beings as vertices and perform specified function through interactions among vertices, such as scientific collaboration network[1, 2], language network [3], spreading network [4, 5], earthquake network [6, 7], stock network [8, 9], evolutionary network [10, 11] and so on. The study of networks, network theory, is an useful tool for analyzing complex system, which is an emerging area of science [12], and has many applications, especially in the interdisciplinary areas [13, 14, 15]. Even though there exist myriads of real networks in the world, scientists are more willing to believe most of networks are marked by several features and governed by countable laws. In recent years, many progresses have drawn continuously increasing attentions and brought us different understandings and new insights [16, 17, 18, 19, 20, 21, 22, 23].

One of the amazing points in the research of complex networks is that the objects in networks are not separated, but interact with each other through links. This interaction can be investigated from the viewpoint of the basic structural feature of complex networks. For example, spectral graph theory studies the properties of graphs through the spectra of their representing matrices [24, 25]. Graph energy is associated with the graph structure in the sense of building cost [26, 27]. Entropy is a general measure of probabilistic uncertainty and proves to be related to the mathematical structure of a complex network [28, 29, 30, 31, 32].

Besides the exploration on the structure of complex networks, the study of dynamics behaviors focuses on the interaction among individuals directly. Individual changes its state through the influence of its neighbors, and also exerts its influence to its neighbours to cause the possible change of their states. For example, opinion dynamics is one of the collective dynamical phenomena in the society [33]. The convergence (or divergence) of opinions among participants of a system is realized through the interactions among participants. Moreover, the interaction phenomenon sometimes does not only happen among individuals, but among independent networks. For example, the failure of nodes in interdependent networks generally not only leads to the failure of their neighbors, but leads to the failure of their dependent nodes in other networks, which may in turn cause further damages to the first network, resulting in cascading failures and catastrophic consequences in the entire system [34].

Additionally, interaction itself is also an interesting object. Take a simple example, in real-world complex systems, many relationships are (relatively) permanent, such as the existing collaboration relationship between two authors in scientific collaboration networks. However, there also exist some relationships which do not persist over long time, but only exist during short periods of time. This temporal effects of complex systems can be taken into account by the temporal networks [35] (also called time-varying graphs [36], or evolving graphs [37]).

2. Spectral graph theory

For a long time, graph theory has played a vital role in analyzing and understanding networks structures. All the topological information of one network can be found in its connectivity matrices. As a related object to matrices, eigenvalue is naturally used to explore the structural properties of a graph. The initial question is that how much information of a graph is contained in its eigenvalues sequence. This problem is described by spectral graph theory, which is the study of properties of a graph in relationship to eigenvalues of its associated matrices.

In the early days, matrix theory were used to analyze the adjacency matrix of graph. There is a large amount of literatures on algebraic aspects of spectral graph theory, such as *Biggs* [38], *Cvetković* [39] and *Godsil et al.* [40]. In a way, spectral graph theory can be considered as the well-developed theory of matrices, of which the purpose is to be related to graph theory and applications with its own characteristics.

In the recent years, many developments in spectral graph theory have the geometric favors on graph, as shown by *Chung* [24] and *Jost* [25]. An important development of spectral graph theory is the interaction with Riemannian geometry. For example, the Cheeger constant from Riemannian geometry has a discrete analogue, which can provide estimation for the first non-zero eigenvalues of the Laplacian matrix [41].

In network theory, besides modeling of complex networks, considerable attention has been given to the problems of capturing topological properties. In particular, it was proved that the important information on the topological properties of a network can be extracted from its spectra [24, 39]. For example, in 1955, Wigner introduced Wigner semicircle law, for certain special classes of random matrices [42, 43]. According to this, the distribution of eigenvalues of a large real symmetric random matrix follows a semicircle distribution. Specifically, the eigenvalues of adjacent matrix and Laplacian matrix of Erdős-Rényi random networks follow the semicircle distributions [44]. The eigenvalues distributions of Watts-Strogatz small-world networks and some of power-law networks are quite far form semicircle [44, 45]. Besides, Banerjee summed up Laplacian spectra for different types of networks and introduced a tentative classification scheme for empirical networks based on their Laplacian spectra [46]. He also revealed more understanding about Laplacian spectra and some particular dynamical processes on networks.

The search on distinguishing the unique characteristics for a given system and uncovering the universal features on large set of systems attracts great interests of scientists [47, 48, 49]. Three generic models support rough classification among networks, but can not provide further classification, especially for networks within the same class. Motif supplies a deep insight into the networks functional abilities associated with evolutionary process, of which analysis implies a similarity measurement based on a comparison of subgraphs frequencies on networks [50]. One problem with this method returns back to the same one shown in graph case, if motifs are too large, then the isomorphism problem appears again. Furthermore, for one important class of networks in our world, biological systems, there are lots of evidences showing the evolutionary process on them can result in their structure changes [51, 52]. The comparison of structures of biological networks may bring us new insights about evolutionary mechanism in biological systems. Without analyzing sequence data of gene or genome, this comparison, focusing on the whole network structure, may support us one way in understanding the functions and evolutions of biological system more conjointly and more intuitively. There is no doubt that graph spectra theory becomes one important and useful tool for analyze networks and brings crucial insights on hiding geometry information of networks.

3. Graph energy

The energy of a graph is an important parameter related to the building cost of the graph. Generally speaking, low energy means low cost, and then, which can be connected to the graph with better structure. Briefly, the energy of a graph is defined as the sum of the absolute values of the eigenvalues of its adjacency matrix. This definition, introduced by *Gutman* in the 1970s [26], is coming from chemistry where it was used to approximate the total π-electron energy of molecules [53]. Initially, graph energy did not draw many attentions of either mathematicians or physicists. Until the coming of the 21st century, extensive research about the graph energy started.

Researches related to the energy of graph can be traced to the 1940s or even 1930s [54, 55, 56, 57]. A finite and undirected graph G with N nodes is denoted to be "hyper-energetic" if its energy is larger than $2(N-1)$ [58], which corresponds to the energy of a complete graph with N vertices. However, if the energy of a connected graph G is less than N, then the graph will be called "hypo-energetic" graph [59]. And another one that corresponds to the "hypo-energetic" graph is the strong graph with energy being smaller than $(N-1)$ [60]. The structure of graph G on N vertices with energy N can be constructed by $N/2$ isolated edges, or $N/4$ isolated quadrangles or something else. In all graphs with N vertices, star graph owns the lowest energy $2\sqrt{N-1}$, except the trivial empty graph whose energy is zero. Actually, the "hypo-energetic" graphs are not so common as *Li* illustrated all this class of graphs with the maximum degree at most 3 [61]. But the "hyper-energetic" graphs are very probable. *Gutman* has found that "hyper-energetic" graphs of order N exist for all $N \geq 8$ [58], which means that there are no "hyper-energetic" graphs with less than 8 vertices. And all graphs with more than $(2N-1)$ edges are "hyper-energetic" [62].

Researchers are striving to find the upper and lower bounds of energy for some special graphs. Firstly, for a graph on N vertices, the energy is less than or equal to $\frac{N}{2}(1+\sqrt{N})$, where the upper bound is related to a strongly regular graph with some special parameters [63], which are equivalent to a certain type of Hadamard matrices. *Haemers* surveyed constructions of the corresponding Hadamard matrices and the related strongly regular graphs [64]. While the energy of a graph with given vertices N and edges m will be less than or equal to $\sqrt{2mN}$ according to the Cauchy-Schwarz inequality with equality holding if and only if the graph is either empty or 1-regular [65]. And for a regular graph with given vertices and average degree, the energy will also have a largest value [66], which is deduced from the energy of a very general graph on N vertices with $m(\geq N/2)$ edges, whose upper bound for the energy is $\frac{2m}{N}+\sqrt{(N-1)[2m-(\frac{2m}{N})^2]}$, where equality holds for some special networks, see Ref. [63]. The results we have mentioned are limited to a general graph. However, most lower or upper bounds are just corresponding to some special graphs, such as the bipartite graph, whose eigenvalues are symmetrical about zero.

The upper bounds for graph energy we have introduced are for the graph with given vertices. However, if a graph is just given the number of edges m but with unknown vertices, the graph energy will also be restricted to a region $[2\sqrt{m}, 2m]$ [67]. The lowest value will be taken if a graph consists of a complete bipartite graph and some arbitrarily isolated vertices. The energy will reach the upper bound while the graph is consisting of m copies of P_2 (a path of two connected vertices) and some isolated vertices. Except for the energy, the largest eigenvalue of adjacency matrix for a general graph will be larger than or equal to the second moment of degree sequence [68]. And this equality will be held only by regular or semiregular bipartite graph. In 2003, *Koolen* and *Moulton* strictly proved that there is a upper bound for the energy of a general bipartite graph according to the number of vertices [69]. So, for all bipartite graphs with the same vertices, there will be a graph with special topological structure corresponding to the maximal energy.

We know that the lower or upper bounds for the graph energy are just connected to some special graphs. That means that the minimal or maximal energy graphs are very limited. However, this limitation is inapplicable for those graphs without the minimal or maximal energy. In 2004, *Balakrishnan* established the existence of equienergetic non-cospectral graphs [66]. He defined that two graphs on the same vertices are called equienergetic if they have the same energy. When two graphs own the same spectra, then they will be called cospectral. Take as a simple example, two simple graphs with two isolated edges and with a quadrangle that are equienergetic non-cospectral with the same energy 4. While the spectra of them are $\{-1, -1, +1, +1\}$, and $\{-2, 0, 0, 2\}$respectively. There is no doubt that two cospectral graphs are also equienergetic.

In recent years, many researches have been studying the graph energy related to the regular graphs [70, 71]. This may be driven from the isotropic of regular graphs that can be easily analyzed from the theoretical view. *Gutman* firstly pointed out that the energy of k-regular graph G on any N-vertices is greater than or equal to N [72]. Therefore, a regular graph with degree $k > 0$ will be never hypo-energetic. If G is also triangle-and quadrangle-free, the minimal value for the energy will depends on k. In Ref. [72], Gutman also showed a remarkable narrow interval for the energy of fullerene or nanotube which can be represented as a 3-regular graph without triangles and quadrangles [73].

4. Opinion dynamics with social diversity

Opinion dynamics is one of the collective dynamical phenomena in society. Convergence (or divergence) of opinions among participants of a debate is a very important social process [74], which is similar to the phase transition from disorder to order of Ising model in statistical physics. Here, we review some works about the effect of social diversity of agents in the opinion dynamics briefly.

In real society, the diversity of agents can be described from different aspects, including the social status, the psychological attitude and mental path. The agents are diverse in their wealth and social status and have diverse influence on others. For simplicity, *Guan et al.* considered the two types of agents A and B with different influence activity in the majority rule model and find that the role of the heterogeneous influence in the order-disorder transition [75]. For example, social leaders have stronger influence compared to normal populations and have a better chance to be followed [76]. And in social networks, the social power of agent may also be quantified as a proportion of its connection degree [76, 77, 78, 79, 80], since the social leader has many followers, namely vertices to which he/she is linked generally. *Kandiah* and *Shepelyansky* introduced the PageRank method to weight the nodes' social power and proposed the PageRank opinion formation (PROF) model [81].

The diversity of agents can also be described according to the psychological attitude and mental path. In real life, people are always rational and make decision through team collaboration or group debate. Then they update their opinion following the rule of peer pressure in majority-rule model. However, some people are inflexible and contrarians, which play an important role in the opinion formation [82]. The inflexible reflects the inertia effect of human during making decision. In contrast to the floater agent who updates its opinion according to the rule of opinion model, inflexible agents keep their opinion always unchanged. *Galam* and *Jacobs* [82] studied the role of inflexible minorities in the democratic opinion model following the local majority firstly. *Biswas* and *Sen* introduced the inflexible in a model of binary opinions in which the updating of agent's opinion according to the state of their neighboring domains with the probability ρ [83]. *Masuda et al.* introduce the parameter ϵ quantifying the strength of the intrinsic preference or partisanship in the voter model [84]. When $\epsilon = 1$, each voter becomes a zealot that never changes opinion (i.e., inflexible) after aligning with its innate preference, when $\epsilon = 0$ reduces to the classic voter model. Here, the zealot effect of voters that never change opinion (i.e., the inflexible effect) has been studied in Refs. [85, 86]. Moreover, the opinion leaders also are considered as the special inflexible [87].

The other psychological attitude is the contrarians, who are the agents that deliberately decides to oppose the prevailing choice of others. *Mobilia* and *Redner* introduced a model of opinion formation according to the majority versus minority with the probability p [88]. *Borghesi* and *Galam* introduced the contrarian effect in the Galam model with a constant density of contrarian a for both opinions to study the chaotic, staggered and polarized dynamics [89]. *Ding et al.* introduced the application of game theory to model the opinion dynamics [90]. Furthermore, the contrarian effect has been introduced in the Sznajd model [91, 92] and q-voter model [93] through a stochastic parameter p. However, those previous works describe the contrarian effect as a constant stochastic parameter, which is too simple to describe the heterogeneous property of agents in social networks. Probably, the contrarian effect can also be determined by the present status of agents, such as the change of its local environment. *Grauwin* and *Jensen* propose a natural, thermal noise which allows for a small probability of interaction between agents when the opinion difference $\Delta > \epsilon$ in Deffuant model with the form of $p_{conv} = [1 + exp((\Delta/\epsilon - 1)/T)]^{-1}$, where T resembles a temperature and characterize the steepness of the convergence and called interaction noise that shows the contrarian effect indirectly [94].

5. The robustness study of interdependent networks

The robustness of interdependent networks has attracted a great deal of attention and understanding how robustness is affected by the interdependence is one of the main challenges faced when designing resilient infrastructures. For example, the robustness of critical infrastructure is one of the most important topics all over the world: specially, different kinds of infrastructure have become more and more interactive under modern technology, like communication and power grid systems, water and food supply systems [95].

In 2010, the seminal model of interdependent networks has defined a one-to-one correspondence between nodes of network A and nodes of networks B [34]. Suppose that two networks have the same number of nodes N. Each node $A_i (i = 1, 2, \ldots, N)$ in network A depends on a functioning node B_i in network B, and if node A_i stops functioning owing to attack or failure, node B_i stops functioning, and vice versa. Based on the generating function formalism and percolation theory, a first-order discontinuous phase transition was found in this model, which is totally different from the second -order continuous phase transition found in isolated networks. In Ref. [96], it was shown that, when the strength of coupling between networks is reduced, the percolation transition becomes second-order transition at a critical coupling strength, which enhanced the robustness of the system. In addition, the vulnerability of the system could be increased by the clustering and assortativity within the network components [97, 98]. And a more realistic case with both strength of coupling and connectivity links between the coupled networks was studied in Ref. [99].

However, the assumption that one node in network A depends on only one node in network B is not valid sometimes. In 2011, Ref. [100] investigated a theoretical framework to study the robustness of two interdependent networks with multiple support dependent relations.

Real interdependent networks are usually not randomly coupled: for example, well-connected ports tend to couple to well-connected airports. So Ref. [101] proposed two inter-similarity measures between the interdependent networks and found that the more inter-similar the entire network is, the more robust the system is [102]. The case in which all pairs of interdependent nodes in both networks have the same degree was studied in Ref. [103]. In the real world, a network is not always attacked randomly. Ref. [104] investigated the robustness of fully and partially interdependent networks under targeted attack, respectively.

As most of real systems are not randomly but spatially embedded, it is reasonable to consider the factor of space limitation. In interdependent lattice networks, Ref. [105] found that there is a change from first to second order phase transition at the critical length of dependency links and Ref. [106] concluded that there is no critical dependency and any small fraction of interdependent nodes leads to an abrupt collapse. Moreover, transport process has been explored in coupled spatial networks in Ref. [107].

There are also some other considerations beyond above ones. A network of networks (NON) is taken into account in Refs. [108, 109] with more realistic consideration that there are more than two interdependent networks in many real systems. Antagonistic interaction [110, 111], autonomous nodes [112] and node-weighted [113] are considered in study of interdependent networks, respectively, which are leading to a better understanding of the effect of dependence between networks on the dynamics of interdependent networks.

6. Temporal networks

Temporal networks, compared to static networks, emphasize on the times *when* and *how long* contact events (edges) are present. The addition of time dimension provides a new sight into the framework of complex network theory. In temporal networks, structural properties and spreading dynamics are constrained by the time ordering of edges. Consequently, the concepts and methods for temporal networks need to be extended or redefined based on the top of static graphs.

In the research of temporal networks, aggregated static networks play a crucial role in understanding the temporal effect of network structure due to the lack of methods to uncover full contact patterns. We can aggregate temporal networks to a list of snapshots of static graphs if the topological characteristics are more relevant than the temporal properties, since it is usually easier to analyze static networks. A temporal network is described as $G(V, E, t, \delta t)$, in which the contact event happens at time t and δt is its duration. An edge is formed in the aggregated network if there is at least one contact happening in time window $[t, t + \Delta t]$. It is noticed that the aggregation time interval Δt has critical consequences on the structural properties emerging from temporal networks. Many existing tools of static graphs have been adopted to analyze temporal networks. For instance, the error and attack vulnerability of temporal networks [114], optimal way for constructing static snapshots in temporal networks [115], and so on.

The adjacency matrix in a temporal network is defined as $a(i, j, t) = 1$ if there exists an edge between node i and node j at time t, and $a(i, j, t) = 0$ otherwise. Given the adjacency matrix as a function of time, the path of temporal networks has two distinct definitions [36]. One corresponds to topological distance, which is analogous to shortest path length in static networks. The other is referred to temporal distance, which means the path of minimum duration to reach each other. Keep these time-respecting paths in mind, we can redefine temporal degree, betweenness, centrality, closeness, component, motif, to name a few. However, not all the structural concepts have their counterparts in static graphs. *Holme* introduced the concept of reachability as a time-ordered chain of contact leading from one node to the other [116]. *Lentz et al.* proposed accessibility to measure temporal networks [117]. See Ref [35] for a review of more details.

In order to find the fundamental role of time ordering and duration, the other direction is to gain insights into the effects of different time correlations. Null temporal models are served as a reference, in which the original time sequences are randomized. It allows us to distinguish between different contributions to the time correlations coming from randomizing contact times, nodes, edges, or combinations of the three.

Besides of this, scientists have also interest in the following question: how will the temporal structures affect dynamical processes on temporal networks, and vice versa? For example, concerning contact events exhibit heterogeneous inter-event time distributions, bursty characters [118] have a strong influence on dynamical processes on temporal networks.

The other effort along this line is to control or to avoid the spread. *Lee et al.* have introduced the concepts of Recent and Weight to investigate the immune strategy [119]. Recent and Weight are specified as the most recent contact and the most often contact, respectively. The research towards spreading dynamics of temporal networks has grown in various aspects ranging from cascades [120] and random walks [121] to synchronization [122], and so on.

Despite the promoting results in temporal networks, this field is still in its infancy and there is not yet a general framework for describing and analyzing it. For example, what is the proper (or characteristic) aggregated time window to reflect network structure over time? Dynamical approaches remain rare in describing spreading processes. Standard models are still lacking for the study of temporal networks. By extending theory and analyzing data to account for temporal networks, we can approach a better understanding of time-stamped complex system.

Acknowledgments
This work was partially supported by the National Natural Science Foundation of China under grant No. 10975057, the Programme of Introducing Talents of Discipline to Universities under Grant No. B08033. J. Jiang wants to thank the support of the National Natural Science Foundation of China under grant No. 11405118. L. Chi wants to thank the support of the Fundamental Research Funds for the Central Universities.

References

[1] Newman M 2001 Scientific collaboration networks. I. Network construction and fundamental results *Phys. Rev.* E **64** 016131.
[2] Hui Z, Cai X, Greneche J and Wang Q A 2011 Structure and collaboration relationship analysis in a scientific collaboration network *Chin. Sci. Bull.* **56** 3702-06.
[3] Deng W, Wang D, Li W and Wang Q A 2011 English and Chinese language frequency time series analysis *Chin. Sci. Bull.* **56** 3717-22.
[4] Pastor-Satorras R and Vespignani A 2001 Epidemic spreading in scale-free networks *Phys. Rev. Lett.* **86** 3200.
[5] Gu J, Gao Z and Li W 2011 Modeling of epidemic spreading with white Gaussian noise *Chin. Sci. Bull.* **56** 3683-88.
[6] Abe S and Suzuki N 2004 Scale-free network of earthquakes *Europhys. Lett.* **65** 581.
[7] Abe S and Szuku N 2012 Universal law for waiting internal time in seismicity and its implication to earthquake network *Europhys. Lett.* **97** 49002.
[8] Eom C, Oh G, Jung W, Hawoong J and Seunghwan K 2009 Topological properties of stock networks based on minimal spanning tree and random matrix theory in financial time series *Physica* A **388** 900-06.
[9] Menezes R and Dionísio A 2011 Globalization and long-run co-movements in the stock market for the G7: An application of VECM under structural breaks *Chin. Sci. Bull.* **56** 3707-16.
[10] Seufert A and Schweitzer F 2007 Aggregate dynamics in an evolutionary network model *Int. J. Mod. Phys.* C **18** 1659-74.
[11] Yang H and Wang B 2011 Universal role of migration in the evolution of cooperation *Chin. Sci. Bull.* **56** 3693-96.

[12] Dorogovtsev N and Mendes F 2013 *Evolution of networks: From biological nets to the Internet and WWW* (Oxford: Oxford University Press).
[13] Newman M 2010 *Networks: an introduction* (Oxford: Oxford University Press).
[14] Cohen R and Havlin S 2010 *Complex networks: structure, robustness and function* (Cambridge: Cambridge University Press).
[15] Davidson E and Levin M 2005 Gene regulatory networks *P Natl. Acad. Sci. USA* **102** 4935.
[16] Manshour P and Montakhab A 2014 Contagion spreading on complex networks with local deterministic dynamics *Communications In Nonlinear Sc.* **19** 2414-22.
[17] Tayurskii D and Lysogorskiy Y 2011 Quantum fluids in nanoporous media-Effects of the confinement and fractal geometry *Chin. Sci. Bull.* **56** 3617-22.
[18] Dörfler F and Bullo F 2014 Synchronization in complex networks of phase oscillators: A survey. *Automatica* **50** 1539-64.
[19] Liu Z 2011 Signal response amplification of scale-free networks *Chin. Sci. Bull.* **56** 3623-29.
[20] Straeten E and Beck C 2011 Skewed superstatistical distributions from a Langevin and Fokker-Planck approach *Chin. Sci. Bull.* **56** 3633-38.
[21] Li W, Luo Y, Wang Y and Cai A 2011 A mean-field Bak-Sneppen model with varying interaction strength *Chin. Sci. Bull.* **56** 3639-42.
[22] Robledo A 2011 Laws of Zipf and Benford, intermittency, and critical fluctuations*Chin. Sci. Bull.* **56** 3643-48.
[23] Liu Z, Guo L and Du J 2011 Nonextensivity and the q-distribution of a relativistic gas under an external electromagnetic field *Chin. Sci. Bull.* **56** 3689-92.
[24] Chung F 1997 *Spectral graph theory* (American Mathematical Soc.)
[25] Jost J and Joy M 2001 Spectral properties and synchronization in coupled map lattices *Phys. Rev.* E **65** 016201.
[26] Gutman I 1978 The energy of a graph *Ber. Math. Statist. Sekt. Forsch. Graz.* **103** 22.
[27] Wang R, Djanteng S and Kaabouchi A 2011 Investigation of an energy nonadditivity for nonextensive systems *Chin. Sci. Bull.* **56** 3661-65.
[28] Tsallis C and Plastino A 1997 Power-law sensitivity to initial conditions-new entropic representation *Chaos, Solitons and Fractals* **8** 885-91.
[29] Jiang J, Wang R, Pezeril M and Wang Q A 2011 Application of varentropy as a measure of probabilistic uncertainty for complex networks *Chin. Sci. Bull.* **56** 3677-82.
[30] Bi Q and Liu J. Exploring non-equilibrium statistical ensembles *Chin. Sci. Bull.* **56** 3654-60.
[31] Ou C and Chen J 2011 Generalized entropies under different probability normalization conditions *Chin. Sci. Bull.* **56** 3649-53.
[32] Zheng L and Li W 2011 Thermoequilibrium statistics for a finite system with energy nonextensivity *Chin. Sci. Bull.* **56** 3666-70.
[33] Chi L 2011 Binary opinion dynamics with noise on random networks *Chin. Sci. Bull.* **56** 3630-32.
[34] Buldyrev V, Parshani R, Paul G, Stanley E and Havlin S 2010 Catastrophic cascade of failures in interdependent networks *Nature* **464** 1025-28.
[35] Holme P and Saramaki J 2012 Temporal networks *Physics Reports* **519** 97-125.
[36] Casteigts A, Flocchini P, Quattrociocchi W and Santoro N 2012 Time-varying graphs and dynamic networks *Int. J. of Parallel Emergent and Distributed Systems* **27** 346-59.
[37] Ferreira A 2004 Building a reference combinatorial model for MANETs *IEEE Network* **18** 24-29.
[38] Biggs N 1993 *Algebraic graph theory* (Cambridge: Cambridge university press).
[39] Cvetkovic M, Doob M and Sachs H 1980 *Spectra of graphs: Theory and application* (New York: Academic press)
[40] Godsil D and Royle G 2001 *Algebraic graph theory* (New York: Springer).
[41] Cheeger J 1970 A lower bound for the smallest eigenvalue of the Laplacian *Problems in analysis* **625** 195-99.
[42] Wigner E 1993 *The Collected Works of Eugene Paul Wigner: Characteristic vectors of bordered matrices with infinite dimensions I* (Berlin Heidelberg: Springer) 524-40.
[43] Wigner E 1993 *The Collected Works of Eugene Paul Wigner: Characteristic vectors of bordered matrices with infinite dimensions II* (Berlin Heidelberg: Springer) 541-45.
[44] Chung F, Lu L and Vu V 2003 Spectra of random graphs with given expected degrees *P Natl. Acad. Sci. USA* **100** 6313-18.
[45] Farkas I, Derényi I, Barabási A and Vicsek T 2001 Spectra of real-world graphs: Beyond the semicircle law *Phys. Rev.* E **64** 026704.
[46] Banerjee A and Jost J 2008 Spectral plot properties: Towards a qualitative classification of networks *Networks and Heterogeneous Media* **3** 395-11.
[47] Erdős P and Rényi A 1959 On random graphs I *Publ. Math. Debrecen* **6** 290-97.
[48] Watts D and Strogatz S 1998 Collective dynamics of small-worldnetworks *Nature* **393** 440-42.

[49] Barabási A and Albert R 1999 Emergence of scaling in random networks *Science* **286** 509-12.
[50] Milo R, Shen-Orr S, Itzkovitz S, Kashtan N, Chklovskii D and Alon U 2002 Network motifs: simple building blocks of complex networks *Science* **298** 824-27.
[51] Ispolatov I, Krapivsky L and Yuryev A 2005 Duplication-divergence model of protein interaction network *Phys. Rev.* E **71** 061911.
[52] Kim J, Kim I, Han S, Bowie J and Kim S 2012 Network rewiring is an important mechanism of gene essentiality change *Scientific reports* **2** 900.
[53] Gutman I 1977 Acyclic systems with extremal Hückel π-electron energy *Theor. Chim. Acta.* **45** 79-87.
[54] Hall G 1955 The bond orders of alternant hydrocarbon molecules *Proc. Roy. Soc.* **229** 251-59.
[55] Coulson C 1940 On the calculation of the energy in unsaturated hydrocarbon molecules. *Proc. Cambridge Phil. Soc.* **36** 201-03.
[56] Ruedenberg K 1961 Quantum mechanics of mobile electrons in conjugated bond systems. III. Topological matrix as generatrix of bond orders *J. Chem. Phys.* **34** 1884-91.
[57] McClelland B 1971 Properties of the latent roots of a matrix: The estimation of π-electron energies. *J. Chem. Phys.* **54** 640-43.
[58] Gutman I 1999 Hyperenergetic molecular graphs *J. Serb. Chem. Soc.* **64** 199-05.
[59] Gutman I and Radenković S 2007 Hypoenergetic molecular graphs *Indian J. Chem.* **46** 1733-36.
[60] So W, Robbiano M, Nair A and Gutman I 2010 Applications of a theorem by Ky Fan in the theory of graph energy *Lin. Algebra Appl.* **432** 2163-69.
[61] Li X and Ma H 2009 All hypoenergetic graphs with maximum degree at most 3 *Lin. Algebra Appl.* **431** 2127-33.
[62] Hou Y and Gutman I 2001 Hyperenergetic line graphs. MATCH Commun *Math. Comput. Chem.* **43** 29-39.
[63] Koolen J 2001 Maximal Energy Graphs *Adv. Appl. Math.* **26** 47-52.
[64] Haemers W 2008 Strongly regular graphs with maximal energy *Lin. Algebra Appl.* **429** 2719-23.
[65] McCelland B 1971 Properties of the latent roots of a matrix: The estimation of π-electron energies *J. Chem. Phys.* **54** 640-43.
[66] Balakrishnan R 2004 The energy of a graph *Lin. Algebra Appl.* **387** 287-95.
[67] Li X, Shi Y and Gutman I 2012 *Graph Energy* (New York: Springer).
[68] Cvetković Cvetkovic D, Doob M and Sachs H 1980 *Spectra of Graphs: Theory and Application* (New York: Academic).
[69] Koolen J H and Moulton V 2003 Maximal Energy Bipartite Graphs. *Graphs and Combinatorics* **19** 131-35.
[70] Alinaghipour F and Ahmadi B 2008 On the energy of complement of regular line graphs *MATCH Commun. Math. Comput. Chem.* **60** 427-34.
[71] Li X, Li Y and Shi Y 2010 Note on the energy of regular graphs *Lin. Algebra Appl.* **432** 1144-46.
[72] Gutman I, Firoozabadi S, delaPeña J and Rada J 2007 On the energy of regular graphs *MATCH Commun. Math. Comput. Chem.* **57** 435-42.
[73] Ecklund P 1996 *Science of Fullerenes and Carbon Nanotubes* (London: Academic Press).
[74] Gandica Y, Castillo-Mussot M, Vázquez G and Rojasc S 2010 Continuous opinion model in small-world directed networks *Physica A* **389** 5864-70.
[75] Guan J, Wu Z and Wang Y 2007 Effects of inhomogeneous influence of individuals on an order-disorder transition in opinion dynamics *Phys. Rev.* E **76** 042102.
[76] Jalili M 2013 Social power and opinion formation in complex networks *Physica A* **392** 959-66.
[77] Guo L, Gu J and Luo Z 2013 How much information is needed to be the majority during the binary-state opinion formation? *Physica A* **392** 4373-79.
[78] Guo L and Cai X 2009 Continuous opinion dynamics in complex networks *Commun. Comput. Phys* **5** 1045-53.
[79] Yang H, Wu Z, Zhou C, Zhou T and Wang B 2009 Effects of social diversity on the emergence of global consensus in opinion dynamics *Phys. Rev.* E **80** 046108.
[80] Yang H and Wang B 2010 Effects of social diversity on the evolutionary game and opinion dynamics *Physics Procedia* **3** 1859-65.
[81] Kandiah V and Shepelyansky D 2012 PageRank model of opinion formation in social networks *Physica A* **391** 5779-5793.
[82] Galam S and Jacobs F 2007 The role of inflexible minorities in the breaking of democratic opinion dynamics *Physica A* **381** 366-76.
[83] Biswas S and Sen P 2009 Model of binary opinion dynamics: coarsening and effect of disorder *Phys. Rev.* E **80** 027101.
[84] Masuda N, Gibert N and Redner S 2010 Heterogeneous voter models *Phys. Rev.* E **82** 010103.
[85] Mobilia M, Petersen A and Redner S 2007 On the role of zealotry in the voter model *J Stat. Mech-Theroy E.* **8** P08029.
[86] Mobilia M 2003 Does a single zealot affect an infinite group of voters? *Phys. Rev. Lett.* **91** 028701.

[87] Ellero A, Fasano G and Sorato A 2013 Stochastic model of agent interaction with opinion leaders *Phys. Rev.* E **87** 042806.
[88] Mobilia M and Redner S 2003 Majority versus minority dynamics: phase transition in an interacting two-state spin system *Phys. Rev.* E **68** 046106.
[89] Borghesi C and Galam S 2006 Chaotic, staggered, and polarized dynamics in opinion forming: the contrarian effect *Phys. Rev.* E **73** 066118.
[90] Ding F, Liu Y, Shen B and Si X 2010 An evolutionary game theory of binary opinion formation *Physica A* **389** 1745-52.
[91] Schneider J 2004 The influence of contrarians and opportunists on the stability of a democracy in the Sznajd model *Int. J. Mod. Phys.* C **15** 659-74.
[92] Lama M, Loopez J and Wio H 2005 Spontaneous emergence of contrarian-like behaviour in an opinoon spreading model *Europhys. Lett.* **72** 851.
[93] Nyczka P, Sznajd-Weron K and Cislo J 2012 Phase transitions in the q-voter model with two types of stochastic driving *Phys. Rev.* E **86** 1011105.
[94] Grauwin D and Jensen P 2012 Opinion group formation and dynamics: Structures that last from nonlasting entities *Phys. Rev.* E **85** 066113.
[95] Rosato V, Issacharoff L, Tiriticco F, Meloni S, De Porcellinis S and Setola R 2008 Modelling interdependent infrastructures using interacting dynamical models *Int. J. Critical Infrastructures* **4** 63-79.
[96] Parshani R, Buldyrev S and Havlin S 2010 Interdependent networks: Reducing the coupling strength leads to a change from a first to second order percolation transition *Phys. Rev. Lett.* **105** 048701.
[97] Huang X, Shao S, Wang H, Buldyrev S and Stanley H 2013 The robustness of interdependent clustered networks *Europhys. Lett.* **101** 18002.
[98] Zhou D, Stanley H, D′Agostino G and Scala A 2012 Assortativity decreases the robustness of interdependent networks *Phys. Rev.* E **86** 066103.
[99] Hu Y, Ksherim B, Cohen R and Havlin S 2011 Percolation in interdependent and interconnected networks: Abrupt change from second to first order transition *Phys. Rev.* E **84** 066116.
[100] Shao J, Buldyrev S, Havlin S and Stanley H 2011 Cascade of failures in coupled network systems with multiple support-dependence relations *Phys. Rev.* E **83** 036116.
[101] Parshani R, Rozenblat C, Ietri D, Ducruet C and Havlin S 2010 Inter-similarity between coupled networks *Europhys. Lett.* **92** 68002.
[102] Gu C, Zou S, Xu X, Ou Y, Jiang Y and He D 2011 Onset of cooperation between layered networks *Phys. Rev.* E **84** 026101.
[103] Buldyrev S, Shere N and Cwilich G 2011 Interdependent networks with identical degrees of mutually dependent nodes *Phys. Rev.* E **83** 016112.
[104] Dong G, Gao J, Tian L, Du R and He Y 2012 Percolation of partially interdependent networks under targeted attack *Phys. Rev.* E **85** 016112.
[105] Li W, Bashan A, Buldyrev S, Stanley H and Havlin S 2012 Cascading failures in interdependent lattice networks: The critical role of the length of dependency links *Phys. Rev. Lett.* **108** 228702.
[106] Bashan A, Berezin Y, Buldyrev S and Havlin S 2013 The extreme vulnerability of interdependent spatially embedded networks *Nature Physics* **9** 667-72.
[107] Morris R and Barthelemy M 2012 Transport on coupled spatial networks *Phys. Rev. Lett.* **109** 128703.
[108] Dong G, Gao J, Du R, Tian L, Stanley H and Havlin S 2013 Robustness of network of networks under targeted attack. *Phys. Rev.* E **87** 052804.
[109] Cellai D, López E, Zhou J, Gleeson J and Bianconi G 2013 Percolation in multiplex networks with overlap *Phys. Rev.* E **88** 052811.
[110] Zhao K and Bianconi G 2013 Percolation on interacting, antagonistic networks. *J Stat. Mech-Theroy. E.* **5** P05005.
[111] Zhao K and Bianconi G 2013 Percolation on interdependent networks with a fraction of antagonistic interactions *J Stat. Phys.* **152** 1069-83.
[112] Schneider C, Yazdani N, Araújo N, Havlin S and Herrmann H 2013 Towards designing robust coupled networks *Scientific reports* **3** 1969.
[113] Wiedermann M, Donges J, Heitzig J and Kurths J 2013 Node-weighted interacting network measures improve the representation of real-world complex systems *Europhys. Lett.* **102** 28007.
[114] Trajanovski S, Scellato S and Leontiadis I 2012 Error and attack vulnerability of temporal networks *Phys. Rev.* E **85** 066105.
[115] Holme P 2013 Epidemiologically optimal static networks from temporal network data *PLOS Comput. Biol.* **9** e1003142.
[116] Holme P 2005 Network reachability of real-world contact sequences *Phys. Rev.* E **71** 046119.
[117] Lentz H, Selhorst T and Sokolov I 2013 Unfolding accessibility provides a macroscopic approach to temporal

networks *Phys. Rev. Lett.* **110** 118701.
[118] Barabási A 2005 The origin of bursts and heavy tails in human dynamics *Nature* **435** 207-211.
[119] Lee S, Rocha L, Liljeros F and Holme P 2012 Exploiting temporal network structures of human interaction to effectively immunize populations *PloS one* **7** e36439.
[120] Karimi F and Holme P 2013 Threshold model of cascades in empirical temporal networks *Physica* A **392** 3476-83.
[121] Starnini M, Baronchelli A, Barrat A and Pastor-Satorras R 2012 Random walks on temporal networks *Phys. Rev.* E **85** 056115.
[122] Fujiwara N, Kurths J and Díaz-Guilera A 2011 Synchronization in networks of mobile oscillators *Phys. Rev.* E **83** 025101.

Diffusion on spatial network

Zi Hui[1,*], Xiaoyue Tang[2], Wei Li[4], Jean-Marc Greneche[3,5] and Qiuping A Wang[3,5]

[1] School of Electrical and Electronic Engineering, Wuhan Polytechnic University, Wuhan 430032, China
[2] School of Mathematics and Computer Science, Wuhan Polytechnic University, Wuhan 430032, China
[3] LUNAM Universite, ISMANS, 44, Ave. Bartholdi, 72000 Le Mans, France
[4] Complexity Science Center, Institute of Particle Physics, Hua-Zhong (Central China) Normal University, Wuhan 430079, China
[5] LUNAM Universite, Universite du Maine, LPEC, UMR CNRS 6087, 72085 Le Mans, France

E-mail: huizi0117@gmail.com

Abstract. In this work, we study the problem of diffusing a product (idea, opinion, disease etc.) among agents on spatial network. The network is constructed by random addition of nodes on the planar. The probability for a previous node to be connected to the new one is inversely proportional to their spatial distance to the power of α. The diffusion rate between two connected nodes is inversely proportional to their spatial distance to the power of β as well. Inspired from the Fick's first law, we introduce the diffusion coefficient to measure the diffusion ability of the spatial network. Using both theoretical analysis and Monte Carlo simulation, we get the fact that the diffusion coefficient always decreases with the increasing of parameter α and β, and the diffusion sub-coefficient follows the power-law of the spatial distance with exponent equals to $-\alpha$-β+2. Since both short-range diffusion and long-range diffusion exist, we use anomalous diffusion method in diffusion process. We get the fact that the slope index δ in anomalous diffusion is always smaller that 1. The diffusion process in our model is sub-diffusion.

1. Introduction

The lastest decade has witnessed an explosive growth of interest in the description of network structure and dynamical properties [1, 2, 3, 4, 5, 6, 7, 8]. In social networks, there are many rumors, opinions, and diseases etc., which often spread gradually through social networks. People always exchange their ideas through face to face chatting, telephone calls, and emails. All these ways are influenced by spatial structure. Previous works have uncovered that people always make calls and sent emails to someone near them [9, 10]. Infectious diseases are spread because of people interaction and traveling, which are constrained by spatial distance as well [11, 12]. Thus, spatial network is a suitable carrier for us to study such problems. Inspired from the Fick's law, whether we can get the diffusion coefficient to measure the spreading ability of spatial network is the aim of this paper.

In recent years, spatial networks have attracted more and more attention. Recent findings reveal that the spatial distance distribution of links follows a power law or exponential [9, 13, 14, 15, 16]. These distributions are quite natural since, for instance, people tend to have their friends and relatives in their neighborhood, transportation networks often favor shorter distance trips, and many communication networks are mainly dominated by short radio ranges

[13]. To model these systems, scientists have proposed spatially constrained networks embedded in one-dimensional or two-dimensional space with a set of links constructed according to spatial condition [17, 18, 19, 20, 21, 22]. What's more, various processes take place on spatial networks, and whose guiding idea focuses on the effects of space. V. Colizza et al. applied the spatio-temporal evolution model to the historical case of the Black Death, which occurred in the 14th century. During these periods, only few traveling paths were available and typical trips were limited to relatively short distances in one day time scale [23]. The main difficulty for the work is estimating the diffusion coefficient D. In that paper, they assumed that virus spread at a velocity of around 160 kms per year, and obtained $D \approx 10^4$ kms^2/year. But D was still an uncertain parameter. Another striking example can be seen in the SARS outbreak in 2003. From the epidemic, scientists thought that pure local spatial diffusion was not a good model anymore and that the global aspect of transportation network needed to be included in the model. They used metapopulation models to discuss the disease spreading in local and with long-range jumps as well and found out that reducing travel is not an efficient strategies for scale-free networks [23]. Hu et al. studied a malware propagation among WIFI routers and Wang et al. discussed the virus spread using Bluetooth and MMS on the spatially constrained networks [24, 25].

In this paper, we analyze this problem in a stylized model of spatial network. The network is fixed on a two-dimensional space and characterized by a set of agents (nodes) and relationships (links) across these agents through which social influence operates. We therefore consider each agent classified as either "active" or "inactive". If an agent has the product (idea, opinion, disease etc.), we mark it as an active node and set state value equals to 1, otherwise it is an inactive node and set state value equals to 0 . Nodes are connected satisfying the social network property, where spatial distance distribution of links follows power law. Both local diffusion and long-range diffusion are discussed in this paper. The diffusion probability of the product is inversely proportional to the spatial distance between active agent and its inactive neighbor. In the previous model, the spatial distance is ignored [4, 5, 7, 8]. This model is an universal model, considering the effect of spatial network structure itself only. Applying the Fick's first law, we define the diffusion coefficient. It is an important parameter indicative of the diffusion mobility. Following this process, we want to reveal the relationship between network structure and its diffusion coefficient. We hope this findings will help us to understand diffusion behavior on social network, such as epidemic spreading, opinion dynamics etc.

We use both theoretical analysis and Monte Carlo simulation method. The numerical results of spatial network are the average of 20 simulations for different realization of networks under the same parameters with the network size of 10000 nodes if not mentioned. The simulation results of diffusion coefficient are the average of at least 1000 different starting configurations, performed on the above 20 simulations. The details of the model are described below.

2. Spatial Driven Network

To work on the spatial network, we first map the network in a two-dimensional space with periodic boundaries. The model are constructed in the following way:

(a) *Initial condition* : The model starts with an initial state ($t = m_0$) of $m_0 + 1$ all-to-all connected nodes on the 1×1 planar and each of them has a randomly two-dimensional coordinate (x, y), $(-0.5 \leq x < 0.5, -0.5 \leq y < 0.5)$.

(b) *Growth* : At every time step, a new node n is randomly located in (x_n, y_n) on the planar.

(c) *Addition of edges* : Following ideas proposed by previous work [19], the new node n connects with m ($m \leq m_0+1$) previous nodes. The previous node i is selected by the probability π_i.

$$\pi_i = \frac{l_{ni}^{-\alpha}}{\sum_j l_{nj}^{-\alpha}} (0 \le \alpha), \tag{1}$$

l_{ni} is the spatial distance between new node n and previous node i, $\sum_j$ is the sum for all the previous nodes. The growing processes repeat step (b) and (c) until the network reaches the desired size. Accordingly, at each step, the number of nodes increases by one, while the number of edges increases by m ($m = m_0 = 4$ in what follows if not mentioned). There are two limiting cases for the present model: when $\alpha = 0$, the network reduces to the random growing process; when $\alpha \to \infty$, the network is close to the model presented by Ozik et al. (OHO model) [18]. As α increases, the new node prefers to connect with nearer nodes. For different m and α, the degree distribution always grows exponentially and is determined by m as follows [22]:

$$p(k) = \frac{1}{m+1}(\frac{m}{m+1})^{k-m}, (k \ge m). \tag{2}$$

3. Diffusion on spatial network

3.1. Spatial Diffusion Mechanism

The spatial network is a suitable frame to discuss the human spreading dynamics. Based on some real social communication networks, we find that the communication rate is inversely proportional to the spatial distance [9, 10]. Arouse from the previous works [12, 26, 27], we define the spatial diffusion mechanism: the diffusion rate λ_{sj} from the active agent s to its inactive neighbor j is inversely proportional to their spatial distance as follows

$$\lambda_{sj} = (\frac{min_{k \in \Omega_s} l_{sk}}{l_{sj}})^{\beta}, \tag{3}$$

where Ω_s is the set of all inactive neighbors of node s. The distance between active agent and its closest neighbor is used to rescale λ_{sj}, changing from 0 to 1. Therefore, for the closest inactive neighbor, the diffusion rate is 1; for other inactive neighbors, the diffusion rate is inversely proportional to the spatial distance from s to them with power β. We assume β is the parameter of diffusion product itself. For smaller β, the product is easy to diffuse; for larger β it is difficult.

3.2. Diffusion Coefficient

Fick's first law relates the diffusive flux to the concentration under the assumption of steady state [28]. It postulates that the flux goes from regions of high concentration to regions of low concentration, with a magnitude that is proportional to the concentration gradient (spatial derivative), which can be written as:

$$J = -D\nabla c, \tag{4}$$

where D is the diffusion coefficient, J is the diffusion flux and c is the concentration. In spreading networks, the product spreads from active agent to inactive one. When N agents are randomly distributed on 1×1 two-dimensional space, the density of nodes per unit area is N. The state density of the agent is equal to its density multiply its state. For an active agent, the state density is $c = N \times 1 = N$. For inactive agent, the state density is $c = N \times 0 = 0$. The product spreads from high state density agent to low state density agent. So we try to use Fick's first law to solve this problem. In order to get the diffusion coefficient, we make two hypothesis: the diffusion coefficient of the network is represented by its average values on different agents; it is unrelated to the number of active agents. Thus, in the following discussion, we only set one agent to active state and calculate the diffusion coefficient in one time step diffusion.

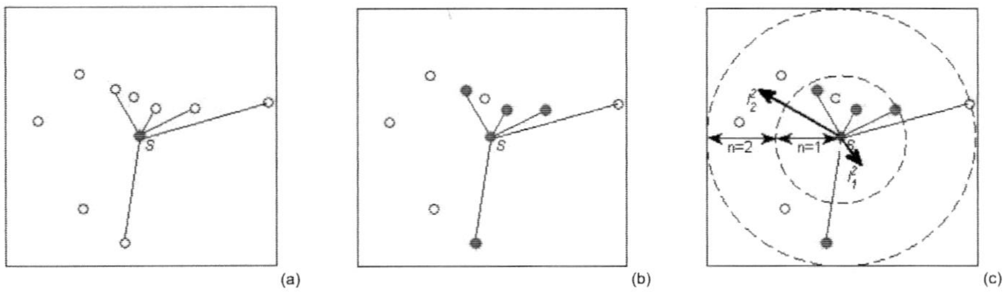

Figure 1. (a) Example of a network and its original state. (b) One step diffusion on the network (a). (c) Divide the network into some rings to calculate the diffusion coefficient in simulation method.

At first, we select an agent s randomly as an active seed, set s to the central of the planar and rescale all the nodes. As Figure 1(a) shows, the red node s in the central represents the active agent, and the circles represent inactive agents. Using spatial diffusion mechanism of Eq. (3) and the Fick's first law of Eq. (4), the diffusion coefficient for node s is given by

$$D_s = \sum_{j \in \Omega_s} \lambda_{sj} \frac{l_{sj}}{N} = \sum_{j \in \Omega_s} \frac{l_{sj}^{-\beta+1}}{(min_{k \in \Omega_s} l_{ks})^{-\beta} N}. \tag{5}$$

The diffusion flux in this model is the number of successful diffusion from seed s. Figure 1(b) is the result of one step spatial diffusion on Figure 1(a).

Then we set agent s as the center and divide the planer to M rings, labeled as $n = 1, 2, \ldots, M$, to calculate D in simulation method (in Fig. 1(c)). The distance between the adjacent two rings is $\frac{1}{2M}$. The distance from the seed to the nth ring is $l_n^M = \frac{n-\frac{1}{2}}{2M}$, which is the average distance from the seed to the two adjacent circles. We assume that the distance between the seed and the nodes in the nth ring has the same value and equals to l_n^M. Since all the agents are randomly located on the planar, the number of agents in the nth ring is $N^M(n) = \frac{\pi(n^2-(n-1)^2)N}{4M^2}$. The approximation diffusion coefficient of the spatial network is the sum of sub-coefficient for all the parts

$$D_s^M = \sum_{n=1}^{M} d^M(n) + d_{out}^M = d^M(1) + d^M(2) + \ldots + d^M(M) + d_{out}^M, \tag{6}$$

d_{out}^M is the diffusion sub-coefficient of the nodes in the planar but out of the nth ring and $l_{out} = \frac{1}{2}$. $d^M(n)$ is the diffusion sub-coefficient contributed from the nodes in the nth ring, satisfying

$$d^M(n) = -\frac{J^M(n)}{\nabla c(n)}. \tag{7}$$

In Figure 1(c), $M = 2$, $J^2(1) = 3$, $J^2(2) = 1$, $N = 10$, $l_1^2 = \frac{1}{8}$ and $l_2^2 = \frac{3}{8}$. According to Eq. (7), $d^2(1) = \frac{3}{80}$, $d^2(2) = \frac{3}{80}$, $D_s^2 = \frac{3}{40}$. When M$\rightarrow \infty$, $D_s^M \rightarrow D_s$.

According to the spatial network evolution mechanism [22], the number of agents, which connect to the seed, in the nth ring is

$$\begin{aligned} N_C^M(n) &\sim N^M(n)(l_n^M)^{-\alpha} \\ &\sim \frac{2\pi N(n-\frac{1}{2})^{-\alpha+1}}{(2M)^{-\alpha+2}}. \end{aligned} \tag{8}$$

Thus, the diffusion flux from the seed to the nth ring satisfies

$$\begin{aligned} J^M(n) &\sim N_C^M(n)(l_n^M)^{-\beta} \\ &\sim \frac{2\pi N(n-\frac{1}{2})^{-\alpha-\beta+1}}{(2M)^{-\alpha-\beta+2}}. \end{aligned} \tag{9}$$

Applying Eq. (4), the diffusion sub-coefficient d of the nth ring is affected by n:

$$d^M(n) \sim (n-\frac{1}{2})^{-\alpha-\beta+2}. \tag{10}$$

For large M, each ring has less than two nodes. We use Eq. (4) to calculate d for each of the nodes, and D_s follows

$$D_s = -\sum_{p\in\Theta_s} \frac{1}{\frac{\Delta c}{l_{sp}}}, \tag{11}$$

where Θ_s is the set of active neighbors, which diffuses from seed s. This process goes for N times, in each of the process, seed s is selected randomly. According to the Ergodic hypothesis, the diffusion coefficient D of the network is the average of D_s.

3.3. Numerical Results

For the traditional Fick's first law, the diffusion coefficient depends on the temperature, viscosity of the fluid and the size of the particles [28]. Now we are looking for which property affects the diffusion coefficient in spatial network. Figure 2 shows the relationship between D and parameter α. We compare three kinds of results for $\beta = 0.5$ and 1: the first one is the simulation result from Eq. (11) on spatial driven network (marked as 'spatial network'); the second one is the simulation result from Eq. (11) on weighted scale-free network, the network grows following scale-free network rule [1] with the same average degree as the first one, and links are weighted by spatial distance from the first one (marked as 'scale-free network'); the third one is the analytical result from Eq. (5) with $M = 100$ on the same spatial driven network (marked as 'analytical result'). The simulation result matches the analytical result. In Figure 2, we also find that the scale-free network and spatial driven network have the same diffusion coefficient. These two networks have the same spatial distance distribution of links but other various properties, such as degree distribution, clustering coefficient, shortest path length. Thus, the diffusion coefficient is determined by the spatial distance distribution of links. D decreases with the increasing of α. We have done it as well in one-dimensional line ($w = 1$), which has the similar relationship. In Figure 3, the left and the lower coordinates correspond to one-dimensional case ($w = 1$), and the right and the upper coordinates correspond to the two-dimensional case ($w = 2$), the dash line satisfies $\alpha = w$. When α is less than the network dimension w, D decreases quickly; when α is greater than the network dimension, it decreases slowly.

β is the parameter of spatial diffusion mechanism corresponding to the property of diffusion particles in Fick's first law. In Figure 4, we discuss how D is affected by parameter β on different spatial driven networks with $\alpha = 0.5$ and 1. For different networks, D always decreases with the

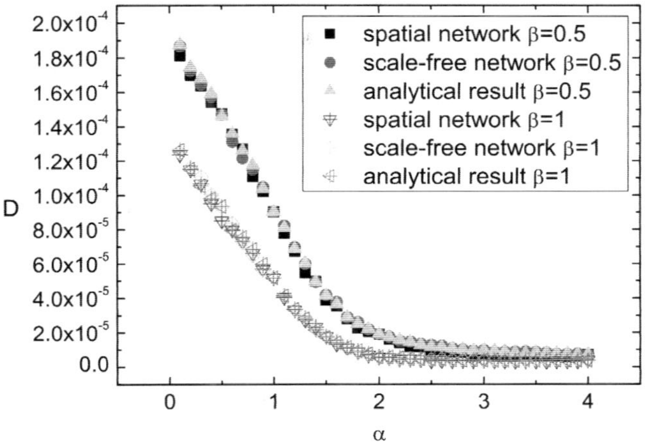

Figure 2. Relation between the diffusion coefficient D and parameter α for $\beta = 0.5, 1$. D is calculated in three different ways.

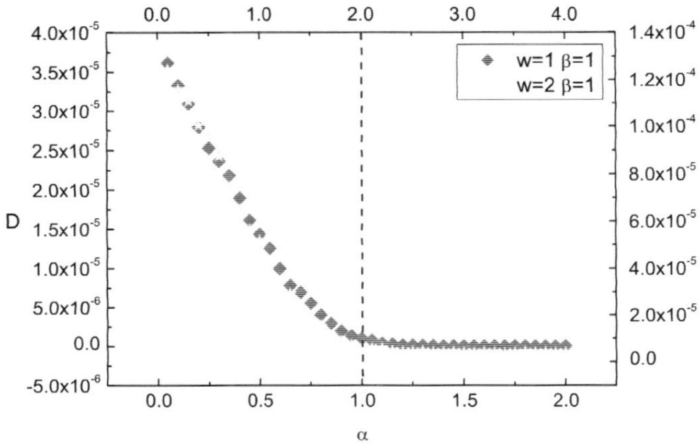

Figure 3. Diffusion coefficient D as a function of parameter α for $w = 1$ and 2 with $\beta = 1$. The left and the lower coordinates correspond to the one-dimension ($w = 1$) line network space, and the right and the upper coordinates correspond to the two-dimension ($w = 2$) plane network space. The dash line follows $\alpha = w$.

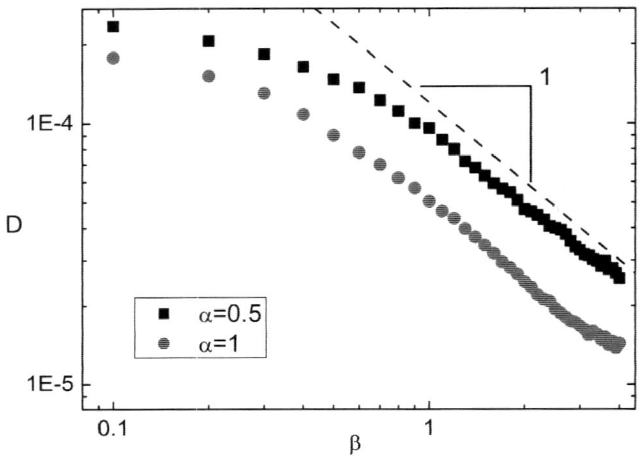

Figure 4. Diffusion coefficient D as a function of parameter β for $\alpha = 0.5$ and 1. The dashed line has a slope -1.

increasing of β, which we have got from Eq. (5) as well. When β is larger than 1, D decreases following the power of β with exponent -1, as Figure 4 shows. In Figure 5, the points suggest that, for $\alpha = 1, 2$ and 3, diffusion sub-coefficient d scales as $d \sim (n - \frac{1}{2})^{\gamma}$. The variation of the exponent γ with $n - \frac{1}{2}$ is shown in Figure 6. The error bars are determined by the fitting with $d \sim (n - \frac{1}{2})^{\gamma}$. The results suggest that approximately

$$\gamma = -\alpha - \beta + 2 \tag{12}$$

Figures 5 and 6 show consistently between the results of numerical simulation and analytical calculation from Eq. (10) for different values of α and β. When $\alpha < -\beta + 2$, long-range connection neighbors are in the main role of diffusion coefficient; when $\alpha > -\beta + 2$, short-range connection neighbors are in the main role.

4. Anomalous diffusion on spatial network

Normal diffusion has as basic characteristic the linear scaling of the mean square displacement of the particles with time, $\langle l^2 \rangle \sim Dt$ [29]. Many different experiments though, including the one shown in the previous section, reveal deviations from normal diffusion, in that diffusion is either faster or slower, and which is termed anomalous diffusion [29, 30]. A useful characterization of the diffusion process is again through the scaling of the mean square displacement with time, where though now we are looking for a more general scaling of the form

$$\langle l^2(t) \rangle \sim t^{\delta}. \tag{13}$$

Diffusion is then classified through the scaling index δ. The case $\delta = 1$ is normal diffusion, all other cases are termed anomalous. The case $\delta > 1$ forms the family of super-diffusive processes, and the case $\delta < 1$ is sub-diffusion process. Since the long-range connections exist in the spatial network, we try to use the anomalous diffusion model to discuss the diffusion process on the

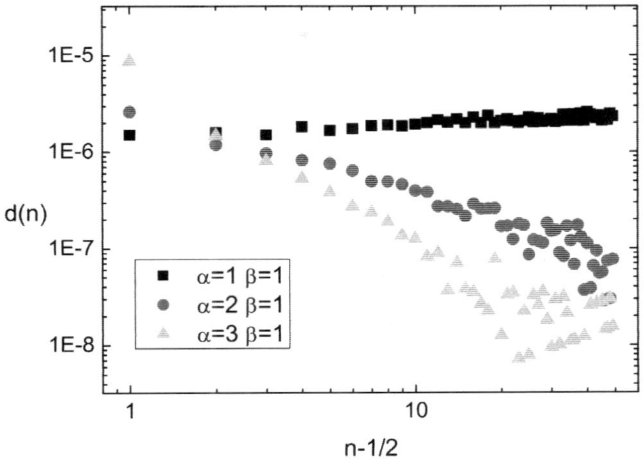

Figure 5. Diffusion sub-coefficient d as a function of $n - \frac{1}{2}$ for $\alpha = 1, 2, 3$ and $\beta = 1$.

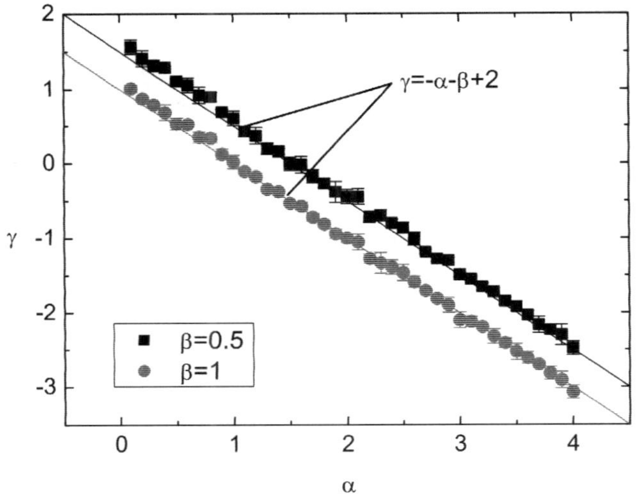

Figure 6. α dependence of the slope γ for $\beta = 0.5, 1$. Points are best fit estimates to the simulation data. The straight lines are plots of Eq. (12).

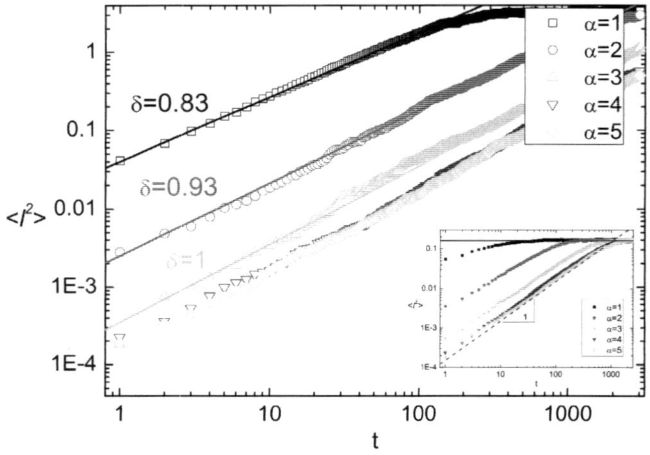

Figure 7. The variation of mean squared displacement for 3×3 expand spatial driven network with different α. inset: The variation of mean squared displacement for 1×1 spatial driven network with different α. The full line satisfies $\langle l^2 \rangle = 0.1667$; the dash line has a slope 1. The nodes represent the ensemble average results for 1000 simulations on a network sample.

spatial network. At first, we select an agent s randomly as an active seed, set s to the central of the planar and rescale all the nodes. The diffusion probability λ_{sj} from the active agent s to its inactive neighbor j is inversely proportional to their spatial distance and follows Eq. (3) as well. At each time step, only one inactive neighbor changes its state. For the active agent, it changes its state at the same time. Thus at every time step, there is only one active agent in the network. At time t, we mark the active agent as s_t. $l(t)$ is the Euclidean distance between active seed s and active agent s_t. The variation of mean squared displacement for different spatial driven network with $\alpha = 1$, 2, 3, 4 and 5 is shown in Figure 7 inset. The active agent diffuses from the active seed constraining by diffusion probability. Since the limited space effect, when diffusion time t is large enough, the product with diffuse to any random place on the planar. The limited value of mean squared displacement calculates as

$$\frac{\int_{-0.5}^{0.5}\int_{-0.5}^{0.5}(x^2+y^2)dxdy}{\int_{-0.5}^{0.5}\int_{-0.5}^{0.5}dxdy} \approx 0.1667. \tag{14}$$

The diffusion process only works before the network approaches the limited space effect. The active time increases with the increasing of α. When $\alpha = 1$, the active time is less than 20 time steps. In order to reduce the limited space effect, we expand the network to 3×3 range with periodic boundary condition. The variation of mean squared displacement for different expand spatial driven network is shown in Figure 7. The mean squared displacement of the active agent is a power law function of time as Eq. (13), whose slope δ is determined by the network parameter α. In normal diffusion process, $\delta = 1$.

The relationship between slope δ and network parameter α is shown in Figure 8. The error bars are determined by the fitting with Eq. (13) in diffusion process. For all the α, slope $\delta \leq 1$. The diffusion process on spatial network with $\alpha < 3$ is sub-diffusion. δ increases with the increasing of α and reaches to 1 (normal diffusion).

In order to reveal the spatial network effect on spatial diffusion, we do realizations of the anomalous diffusion on different spatial networks with $\alpha = 1$ and 4. Figure 9 shows the

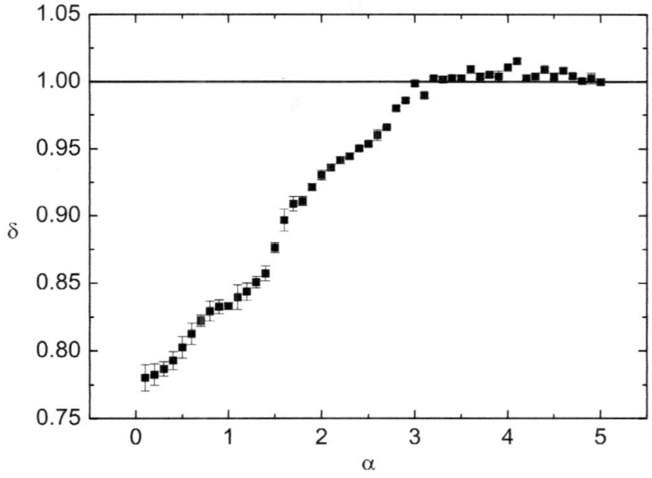

Figure 8. The slope δ dependence of the parameter α. The straight line satisfies $\delta = 1$ for normal diffusion.

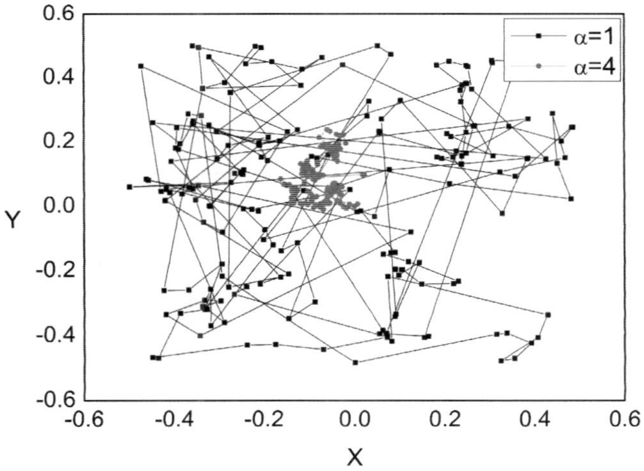

Figure 9. Two trajectories of anomalous diffusion on spatial network (a) $\alpha = 1$ (b) $\alpha = 4$ with $N = 10000$. Both trajectories are simulation results for 100 time steps.

comparison of 100 steps simulation trajectories of anomalous diffusion on spatial network with $\alpha = 1$ and spatial network with $\alpha = 4$. When $\alpha = 1$, the network structure is beneficial for diffusion, existing a lot of long-range diffusion. In 100 time steps, the product has diffused on the whole planar. On the contrary, when $\alpha = 4$, it is normal diffusion. The network structure is disadvantage for diffusion, only short-range diffusions are existed. The product diffuses very slowly.

5. Summary

To summarize, we use the Fick's first law and anomalous diffusion to solve the diffusion dynamics on spatial networks. In the method of Fick's first law, we introduce a composite diffusion coefficient D in order to measure the diffusion ability of the spatial driven networks. Since our diffusion is not normal in general due to the long-range spreading, this coefficient must take into account the transfers of diffusive substances over all distances. We get diffusion coefficient D using both of theoretical method and simulation method and find that D is determined by spatial distribution of links. In simulation method, we divide the planer to M rings, and apply the Fick's first law on each ring to get sub-coefficient d. For larger M, according to the simulation method, D is the sum of diffusion sub-coefficient for all the rings. When $\alpha < w$, where w is the topological dimension of the space, the diffusion coefficient decreases quickly with the increasing of α. On the contrary, it decreases slowly. For different diffusion mechanism, D always decreases with the increasing of β, where β is the parameter of diffusion mechanism. When $\beta > 1$, D decreases following the power of β with exponent -1. Finally, the diffusion sub-coefficient $d(n)$ is determined by the network structure and position of the ring, which follows $d(n) \sim (n - \frac{1}{2})^{-\alpha-\beta+2}$, where n is the label of the ring. In the method of anomalous diffusion, the diffusion probability is inversely with the Euclidean distance between two connected agents as well. From the variation of mean squared displacement, we get that the slope index δ in anomalous diffusion increases with the increasing of α and then reaches to 1. The diffusion process in our model is always sub-diffusion.

In anomalous diffusion, the diffusion process is characterized through the scaling of the mean square displacement with time. This is a mature method to discuss the diffusion process. But it works only on the traditional anomalous diffusion process. Once the scaling property of the mean square displacement with time does not exist, the anomalous diffusion method does not work. In this case, the method of Fick's law is a good choice to discuss the diffusion process. In section 3.1, the diffusion dynamics works on more than one pair of agents. In this condition, the mean square displacement does not easy to get. Thus, using the Fick's law on all pairs of agents is a more suitable method. At this point we must emphasize that our approach represents a attempt to use the Fick's first law and anomalous diffusion on complex network. For the future, we will try to apply the diffusion coefficient on different network dynamic processes, such as opinion dynamic (Voter model, Sznajd model), spreading dynamic (SIS model, SIR model) et al.

Acknowledgments

This work was Supported by the Talent Scientific Research Foundation of Wuhan Polytechnic University (Grants No.2014RZ09 and No.2014RZ06), and the PHC CAI YUAN PEI Programme (LIU JIN OU [2010] No. 6050).

References

[1] Barabási A L and Albert R 1999 *Science* **286** 509
[2] Watts D J and Strogatz S H 1998 *Nature* **393** 440
[3] Donetti L, Hurtado P I and Muñoz M A 2005 *Phys. Rev. Lett.* **95** 188701
[4] Sznajd-Weron K 2005 *Acta Phys. Pol.* B **36** 2537
[5] Sznajd-Weron K and Sznajd J 2000 *Int. J. Mod. Phys.* C **11** 1157
[6] Santos F C, Roderigues J F and Pacheco J M 2005 *Phys. Rev.* E **72** 056128
[7] Pastor-Satorras R and Vespignani A 2001 *Phys. Rev.* E **63** 066117
[8] Pastor-Satorras R and Vespignani A 2002 *Phys. Rev.* E **65** 036104
[9] Lambiotte R, Blondel V D, Kerchove de C, Huens E, Prieur C, Smoreda Z and Dooren Van P 2008 *Physica* A **387** 5317
[10] Goldenberg J and Levy M 2009 *arXiv*: 0906.3202
[11] Barrat A, Barthélemy M, Pastor-Satorras R and Vespignani A 2004 *Proc. Natl. Acad. Sci.* **101** 3747
[12] Hui Z, Cai X, Greneche J M, Wang Q A 2012 *Int. J. Mod. Phys.* C **23** 1250082

[13] Barrat A, Barthelemy M, and Vespignani A 2005 *J. Stat. Mech.* P05003
[14] Liben-Nowell D, Novak J, Kumar R, Raghavan P and Tomkins A 2005 *Proc. Natl. Acad. Sci.* **102** 11623-8
[15] Yook S -H, Jeong H and Barabási A -L 2002 *Proc. Natl. Acad. Sci.* **99** 13382
[16] Hayashi Y 2006 *IPSJ Trans. Special Issue on Network Ecology* **47 (3)** 776-85
[17] Sen P and Manna S S 2003 *Phys. Rev.* E **68** 026104
[18] Ozik J, Hunt B R and Ott E 2004 *Phys. Rev.* E **69** 026108
[19] Kosmidis K, Havlin S and Bunde A 2008 *Europhys. Lett.* **82** 48005
[20] Manna S S and Sen P 2002 *Phys. Rev.* E **66** 066114
[21] Xulvi-Brunet R and Sokolov I M 2002 *Phys. Rev.* E **66** 026118
[22] Hui Z, Li W, Cai X, Greneche J -M and Wang Q A 2013 *Physica* A **392** 1909
[23] Colizza V, Barthelemy M, Barrat A and Vespignani A 2007 C. R. *Biologies* **330** 364
[24] Hu H and Myers S, Colizza V and Vespignani A 2009 *Proc. Natl. Acad. Sci.* **106** 1318
[25] Wang P, Gonzalez M, Hidalgo C A, and Barabasi A -L 2009 *Science* **324** 1071
[26] Sun P, Cao X B, Du W B and Chen C L 2010 *Phys. Pro.* **3** 1811
[27] Guo W P, Li X and Wang X F 2007 *Physica* A **380** 684
[28] Crank J 1975 *The mathematics of diffusion* 2ed, Oxford
[29] Vlahos L, Isliker H, Kominis Y and Hizanidis K 2008 *arXiv*: 0805.0419
[30] Metzler R and Klafter J 2000 *Physics Reports.* **339** 1

Proteins aggregation and human diseases

Chin-Kun Hu
Institute of Physics of Academia Sinica, Taipei 11529, Taiwan
E-mail: huck@phys.sinica.edu.tw

Abstract. Many human diseases and the death of most supercentenarians are related to protein aggregation. Neurodegenerative diseases include Alzheimer's disease (AD), Huntington's disease (HD), Parkinson's disease (PD), frontotemporal lobar degeneration, etc. Such diseases are due to progressive loss of structure or function of neurons caused by protein aggregation. For example, AD is considered to be related to aggregation of Aβ40 (peptide with 40 amino acids) and Aβ42 (peptide with 42 amino acids) and HD is considered to be related to aggregation of polyQ (polyglutamine) peptides. In this paper, we briefly review our recent discovery of key factors for protein aggregation. We used a lattice model to study the aggregation rates of proteins and found that the probability for a protein sequence to appear in the conformation of the aggregated state can be used to determine the temperature at which proteins can aggregate most quickly. We used molecular dynamics and simple models of polymer chains to study relaxation and aggregation of proteins under various conditions and found that when the bending-angle dependent and torsion-angle dependent interactions are zero or very small, then protein chains tend to aggregate at lower temperatures. All atom models were used to identify a key peptide chain for the aggregation of insulin chains and to find that two polyQ chains prefer anti-parallel conformation. It is pointed out that in many cases, protein aggregation does not result from protein mis-folding. A potential drug from Chinese medicine was found for Alzheimer's disease.

1. Introduction

Up to 2014, there are 8 Nobel Prize winners in sciences with Chinese background, including Chen-Ning Yang (physics prize in 1957, born in 1922 in Hefei, China), Tsung-Dao Lee (physics prize in 1957, born in 1926 in Shanghai, China), Samuel Chao Chung Ting (physics prize in 1976, born on 27 January 1936 in USA), Yuan-Tseh Lee (chemistry prize in 1986, born on 19 November 1936 in Hsinchu, Taiwan), Steven Chu (physics prize in 1997, born on 28 February 1948 in St. Louis, MO, USA), Daniel Chee Tsui (physics prize in 1998, born on 28 February 1939 in Henan, China), Roger Yonchien Tsien (chemistry prize in 2008, born on 1 February 1952 in New York, USA), and Sir Charles Kuen Kao (physics prize in 2009, born on 4 November 1933 in Shanghai, China). Among these 8 Nobel Prize winners, Sir Charles Kuen Kao was identified to have Alzheimer's disease (AD) in early 2003. In tradition, every Nobel Prize Winner should give a Nobel lecture. In 2009, Dr. Kao already lost the ability to present a Noel lecture. Dr. Kao's friends and family members helped him to prepare and present the Nobel lecture. Professor Steven Chu's father Ju-Chin Chu (14 December 1919 - 15 November 2000) was an academician of Academia Sinica in Taipei and had AD. Thus among 8 Nobel Prize winners in sciences with Chinese background, two families had been identified with AD. It is quite a high percentage. It has been estimated that 45% of USA citizens with age higher than 85 have AD. AD and other human diseases become more serious in longevity societies.

Alois Alzheimer (14 June 1864 - 19 December 1915) was a German psychiatrist and neuropathologist. In 1901, he worked at Frankfurt Asylum and found that Mrs. Auguste Deter, 51-year-old patient at the Frankfurt Asylum, had strange behavioral symptoms, including a loss of short-term memory. In 1903, Alzheimer moved to Munich to work at Anatomical Laboratory at the Royal Psychiatric Clinic of the University of Munich. In April 1906, Mrs Deter died and her brain was sent to Alzheimer's lab. in Munich. Alzheimer and two Italian technicians used the staining techniques to identify amyloid plaques and neurofibrillary tangles in Mrs. Auguste Deter's brain. On 3 November 1906, Alzheimer reported this case study at the 37th meeting of the Society of Southwest German Psychiatrists in Tubingen, Germany the clinical and neuropathological characteristics of the disease [1, 2]. In 1910, Emil Kraepelin called the disease identified by Alois Alzheimer as Alzheimer disease (AD) [3].

The 3 November 2006 issue of Science magazine used AD as the cover story and contains two review papers about AD: (1) "A Century of Alzheimer's Disease" by Michel Goedert and Maria Grazia [4], (2) "100 Years and Counting: Prospects for Defeating Alzheimer's Disease" by Erik D. Roberson and Lennart Mucke [5]. It is widely believed that AD is mainly due to aggregation of Aβ40 [6] and Aβ42 [7] in AD patient's brain and such Aβ40 and Aβ42 form hairpin structures [6, 7].

On 15 February 1872, George Huntington (9 April 1850 - 3 March 1916) read before the Meigs and Mason Academy of Medicine at Middleport, Ohio, USA a paper "On Chorea". In the paper, he wrote: "Chorea is essentially a disease of the nervous system. The name chorea is given to the disease on account of the dancing propensities of those who are affected by it, and it is a very appropriate designation. The disease, as it is commonly seen, is by no means a dangerous or serious affection, however distressing it may be to the one suffering from it, or to his friends. Its most marked and characteristic feature is a clonic spasm affecting the voluntary muscles" [8]. The disease is then called "Huntington diseases (HD)" [9].

HD, spinobulbar muscular atrophy, dentatorubral-pallidoluysian atrophy, and six kinds of spinocerebellar ataxias are well known as trinucleotide repeat disorder of inherited neurodegenerations. Theses diseases are considered to be caused by abnormal polyglutamine (polyQ) repeat encoded by expanded CAG repeats in genomic sequence [10, 11, 12, 13, 14]. HD is the most common among these diseases which has prevalence estimated about 4 in 10,000 people of Western European ancestry [15].

In 1817, James Parkinson (11 April 1755 - 21 December 1824) published *An Essay on the Shaking Palsy* [16]. The Essay contains five chapters. In Chapter 1, Parkinson wrote: "Involuntary tremulous motion, with lessened muscular power, in parts not in action and even when supported; with a propensity to bend the trunk forwards, and to pass from a walking to a running pace: the senses and intellects being uninjured. The term Shaking Palsy has been vaguely employed by medical writers in general. By some it has been used to designate ordinary cases of Palsy, in which some slight tremblings have occurred; whilst by others it has been applied to certain anomalous affections, not belonging to Palsy." The disease described by James Parkinson in [16] was later known as Parkinson disease (PD) [17, 18]. It has been proposed that PD can be related to aggregation of α-synuclein: the protein which has 140 amino acids, and is encoded by the SNCA gene [18].

A supercentenarian (or written as super-centenarian) is someone who has lived to or passed his/her 110th birthday. This age is achieved by about one in 1,000 centenarians (lived to or passed his/her 100th birthday). Senile systemic amyloidosis was determined to be the primary cause of death for 70% of supercentenarians who have been autopsied [19].

Thus it is important to understand the mechanism of protein aggregation. In the present paper, we briefly review the application of statistical physics to understand protein aggregation related to human diseases.

The basic content of statistical physics was formulated from late 19th century to early 20th century by Ludwig Eduard Boltzmann (20 February 1844 - 5 September, 1906) [20] and Josiah Willard Gibbs (11 February 1839 - 28 April 1903) [21]. The objective of statistical physics is to understand the properties of a macroscopic system from interactions of the molecules or atoms of the macroscopic system. One important research subfield of statistical physics is the study of phase transitions and critical phenomena [22], for a recent review please read [23]. Universality and scaling are two key concepts developed from the study of phase transitions and critical phenomena [22, 23]. In 1971, Ken G. Wilson developed renormalization group (RG) theory [24, 25, 26] for critical phenomena, which provides a framework for understanding universality and scaling in critical systems, and can be used to calculate critical quantities of critical systems. For this contribution, Ken G. Wilson won the Nobel Prize in physics in 1982 and his Nobel Prize Lecture "The renormalization group and critical phenomena" was published in [27].

In the development of statistical physics, it has often been found that a simple model system can be useful for understanding collective or critical behavior [22, 23] of a complicated system consisting of a large number of atoms or molecules. For example, in 1945 Guggenheim [28] reported that in the T/T_c versus $\rho/\ \rho_c$ plane (T, T_c, ρ and ρ_c are temperature, critical temperature, density, and critical density of the substance, respectively) the coexistence curves of N_e, A_r, K_r, X_e, N_2, O_2, CO, and CH_4 are very consistent with each other, and near the critical point the order parameter [22] $|\rho - \rho_c| \sim |1 - (T/T_c)|^\beta$, with the critical exponent $\beta = 1/3$. In 1952, Yang and Lee proposed that the critical behavior of the gas-liquid system can be represented by a lattice-gas model [29], which is equivalent to the Ising model [30], in which each atom or molecule on a lattice site is assigned a variable which can be either +1 or -1. In 1995-1996, Blöte and collaborators [31, 32] used Monte Carlo simulations to find that the critical exponent β of the spontaneous magnetization and ν of the correlation length [22] of a three-dimensional (3D) Ising model are 0.3269(6) [32] and 0.6301(8) [31], respectively. In 2009, Sengers and Shanks [33] reviewed the experimental data for liquid-gas critical systems and reported that the order parameter and the correlation length have critical exponents $\beta = 0.3245$ and $\nu = 0.629 \pm 0.003$, respectively. In 2012, Watanabe, Ito and Hu [34] used molecular dynamics simulations to find that β and ν of a 3D Lennard-Jones (L-J) model system [35, 36] are 0.3285(7) and 0.63(4), respectively. The values of β and ν obtained for simple model systems reported in Refs. [31, 32, 34] are highly consistent with experimental data reported in [28, 33]. Using Monte Carlo methods [37, 38, 39, 40] and analytic methods [41, 42] it has been found that many critical systems can have very nice universal and scaling behaviors [43, 44, 45, 46, 47, 48, 49, 50].

On the basis of molecular biology, it is of interest to know whether we can apply ideas and methods of statistical physics to understand protein aggregation related to human diseases. The purpose of this paper is to address this problem. In the next section, we first review the usage of lattice models to understand some factors which influence protein aggregation. In section 3, we review the application of an off-model of polymer chains to understand the influence of bending-angle dependent and torsion-angle dependent interactions on the aggregation of proteins. In section 4, we review some developments in the application of all-atom models to understand some problems related to protein aggregation. In section 5, we report the discovery of a potential drug for AD. Finally, in section 6, we summarize and discuss our result.

2. Lattice models for aggregation of proteins

Lattice Ising model has been used to describe critical phenomena of gas-liquid systems very successfully [23, 31, 32, 33, 51]. Thus one would like to use simple lattice models to understand folding and aggregation of proteins, which are polymer. As the temperature of the dilute polymer solution decreases, the polymer will undergo a collapse transition [52, 53]. The interacting self-avoiding walk (ISAW) [54, 55, 56, 57, 58, 59] on the simple cubic (SC) lattice is the simplest lattice model for homogenous polymers in dilute solution. The model has a collapse transition at a higher temperature and a freezing transition at a low temperature.

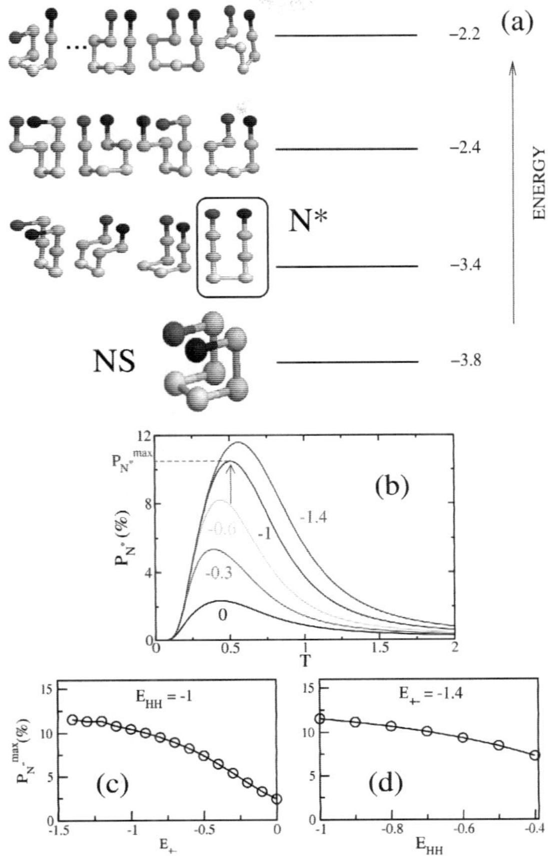

Figure 1. (Color online) (a) Spectrum of energies and associated low energy structures of the monomer sequence +HHPPHH-. H, P, + and - are in green, yellow, blue, and red, respectively. We set $E_{HH} = -1$ and $E_{+-} = -1.4$. There are a total 1831 possible conformations that are spread among 17 possible energy values. The conformations in the first excited state represent the ensemble of **N*** structures and the **N*** conformation that coincides with the peptide state in the fibril [see Figure 2(a)] is enclosed in a black colored box. (b) The probability P_{N^*} of populating the structure in the box in (a) as a function of T for $E_{+-} = 0, -0.3, 0.6, -1$ and -1.4 keeping $E_{HH} = -1$. The arrow indicates T^*, where $P_{N^*} = P_{N^*}^{max}$. Dependence of $P_{N^*}^{max}$ on E_{+-} for $E_{HH} = -1$ (c), and on E_{HH} for $E_{+-} = -1.4$ (d). Reproduced from Fig. 1 of [60].

Chen, Hsieh, and Hu (CHH) [59] developed a novel method to analyze phase behavior of lattice models based on partition function zeros by considering both loci of partition function zeros and thermodynamical functions associated with them. With this method, the first pair of complex conjugate zeros (first zeros) can be defined without ambiguity and the critical point of a small system can be defined as the peak position of the heat capacity component associated with the first zeros. For the system with two phase transitions, two pairs of first zeros corresponding to two phase transitions can be identified and two overlapping phase transitions can be well separated. This method was applied to the interacting self-avoiding walk (ISAW) of homopolymer with N monomers on the simple cubic lattice. The exact partition functions

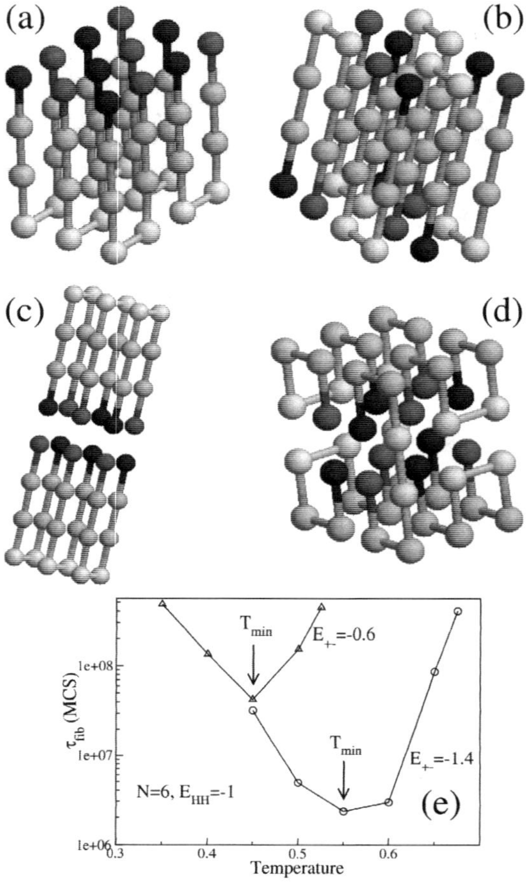

Figure 2. (Color online) (a) The lowest energy fibril structure for $E_{+-} = -1.4$ and $E_{HH} = -1$. (b) Same as in (a) but with $E_{+-} = 0$. (c) Double layer structure for $E_{HH} = -0.4$ but keeping $E_{+-} = -1.4$. (d) For $E_{+-} = -1.4$ and $E_{HH} = -0.3$ the fibril structure is entirely altered. (e) Temperature dependence of τ_{fib} for $E_{+-} = -1.4$ (circles) and $E_{+-} = -0.6$ (triangles). $N = 6$ and $E_{HH} = -1$. Arrows show the temperatures at which the fibril formation is fastest. Reproduced from Fig. 2 of [60].

Z_N with N up to 27 were calculated and CHH approach gives a clear scenario for the collapse and the freezing transitions [59]. CHH's conformation number for the ISAW on the $3 \times 3 \times 3$ SC lattice is consistent with that of [54], which only studied the ISAW on the $3 \times 3 \times 3$ SC lattice.

In 2010, Li *et al.* studied a lattice model for protein aggregation [60]. In this model, an amino acid is represented by a bead on a site of a simple cubic lattice. The bead can be H (hydrophobic), P (polar), +1 (positively charged), or -1 (negatively charged) with interaction parameters: $E(PP) = E(P+) = E(P-) = -0.2$, $E(PH) = E(H+) = E(H-) = 0.2$, $E(+-)$ and $E(HH)$ are varying parameters, $E(++) = E(--) = -E(+-)/2$. $E(+-)$ is between -1.4 and 0 and $E(HH)$ is between -1 and 0. Each chain has $M = 8$ beads and there are $4^M = 65536$ different chains. Among such chains, 5950 chains have a unique ground state and are protein-like. For $E(+-) = -1.4$ and $E(HH) = -1$, the ground state, the first, the second, and the third

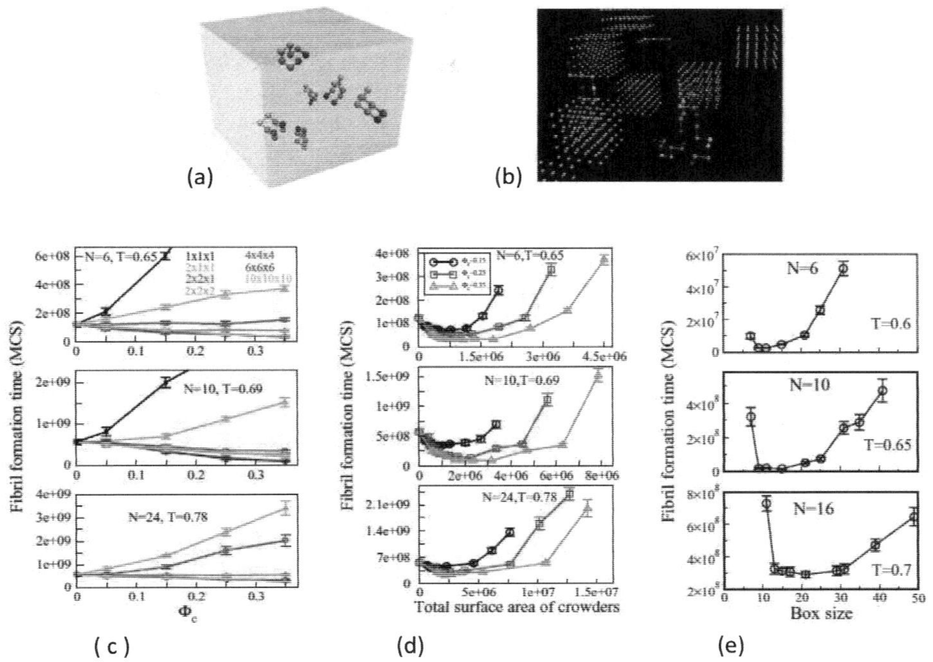

Figure 3. (Color online) (a) Random configuration of chains in a confined box. (b) Snapshot of crowders (grey cubes) and polypeptide chains (color). (c) Dependence of τ_{fib} on the crowder concentration ϕ_c for various sizes of crowders. Orange and red refer to non-cubic (rectangular parallelepiped) crowders with dimensions $2 \times 1 \times 1$ and $2 \times 2 \times 1$. Other crowder sizes are also shown. The number of polypeptide chains N = 6, 10, and 24. (d) Dependence of τ_{fib} on the total surface area of crowders at three values of crowder concentration ϕ_c. T = 0.65, 0.69, and 0.78 for the number of chains N = 6, 10, and 24, respectively. The results are averaged over 50 to 200 MD trajectories for each set of parameters. (e) The U-shape dependence of τ_{fib} on the sizes of confining cubic box. The results obtained from 50 to 200 MC trajectories for each box size. Error bars come from averaging over MC runs. Edited from figures of [62].

excited states are shown in Figure 1(a). When 6 peptide chains were used to simulate aggregation of such chains, one obtained stable conformations of Figures 2(a), 2(b), 2(c), and 2(d) with parameters (a) $E(+-) = -1.4$ and $E(HH) = -1$, (b) $E(+-) = 0$ and $E(HH) = -1$, (c) $E(+-) = -1.4$ and $E(HH) = -0.4$, and (d) $E(+-) = -1.4$ and $E(HH) = -0.3$, respectively. Please note that for $E(+-) = -1.4$ and $E(HH) = -1$, the system used the hairpin structure of the first excited state in Figure 1(a), called N^*, to form the aggregated state of Figure 2(a).

The number of conformations of a sequence of $M = 8$ beads, that satisfy self-avoidance is only 1831. The probability P_{N^*} of populating the structure in the box in Figure 1(a) as a function of T for E_{+-}=0, -0.3, -0.6, -1, and -1.4 keeping $E_{HH} = -1$ had been calculated exactly and plotted in Fig. 1(b), in which arrow indicates T^*, where $P_{N^*} = P_{N^*}^{max}$. Dependence of $P_{N^*}^{max}$ on E_{+-} for $E_{HH} = -1$, and on E_{HH} for $E_{+-} = -1.4$ were shown in Figures 1(c) and 1(d), respectively.

Monte Carlo (MC) method was used to calculate the temperature dependence of the fibril formation time $\tau_{\rm fib}$ for $E_{+-} = -1.4$ (circles) and $E_{+-} = -0.6$ (triangles) with $N = 6$ and $E_{HH} = -1$, and calculated results were shown in Figure 2(e), in which arrows show the temperatures at which the fibril formation is fastest. Figure 2(b) shows that for E_{+-}=-1.4 and $E_{HH} = -1$, P_{N^*} has a maximum at $T = 0.55$, at the same temperature the fibril formation is fastest with the same parameters E_{+-}=-1.4 and $E_{HH} = -1$ in Figure 2(e). Thus Li *et al.* [60] found that the probability of a protein sequence appearing in an aggregated conformation can be used to determine the temperature at which the protein can aggregate most quickly. They also found a correlation between the time taken for the protein to aggregate and the strength of interactions between charged amino acids, which is consistent with previous experimental observations [61].

Co, Hu, and Li used the model of [60] to develop a lattice model for studying the aggregation of polypeptide chains in the presence of crowders [62]. Figure 3(a) shows an example of random conformations of peptide chains in a confined box. Figure 3(b) sows a snapshot of crowders (grey cubes) and polypeptide chains (color). The peptide chains can not enter the lattice sites occupied by crowders and there is no other interactions between peptide chains and crowders. Monte Carlo method was used to study the influence of various factors on the fibril formation time $\tau_{\rm fib}$. One MC step consists of 9 local moves of peptide chains and one global move of a chain; 5 MC moves follow by one global move of all crowders.

Figure 3(c) shows the dependence of $\tau_{\rm fib}$ on the crowder concentration ϕ_c for various sizes of crowders. Orange and red refer to non-cubic (rectangular parallelepiped) crowders with dimensions $2 \times 1 \times 1$ and $2 \times 2 \times 1$. Other crowder sizes are also shown. The numbers of polypeptide chains N are 6, 10, and 24 for top, middle, and bottom panels, respectively. $\tau_{\rm fib}$ of Fig. 3(c) decreases with the concentration of crowders if crowder sizes are large enough, but the growth is observed for crowders of small sizes. Figure 3(d) shows the dependence of $\tau_{\rm fib}$ on the total surface area of crowders at three values of crowder concentration ϕ_c with $T = 0.65$, 0.69, and 0.78 for the number of chains $N = 6$, 10, and 24, respectively. The results are averaged over 50 to 200 MD trajectories for each set of parameters. Such results can be used to explain the recent experimental observation on the dual effect of crowding particles on fibril growth of proteins that for a fixed crowder concentration the fibrillation kinetics is fastest at intermediate values of total surface of crowders [63]. It becomes slow at either small or large coverage of cosolutes.

Figure 3(e) shows the U-shape dependence of $\tau_{\rm fib}$ on the sizes of confining cubic box. The results obtained from 50 to 200 MC trajectories for each box size. Error bars come from averaging over MC runs. Due to competition between the energetics and entropic effects, the dependence of $\tau_{\rm fib}$ on the size of confined space is described by a parabolic function [62].

3. Relaxation and aggregation of polymer chains

Besides protein aggregation mentioned in Section 1, there is another interesting biological problem: How a biological system can maintain in non-equilibrium state? In particular, why ancient seeds can maintain in a viable non-equilibrium state for about 2000 years [64] or even about 30,000 years [65]? This section will review some clues for answering these questions [51].

Many important biological macromolecules, such as DNA, RNA, proteins, etc, are polymers. To understand why a biological system can maintain in a non-equilibrium state for a very long time and why proteins aggregate, one can try to get relevant information from the relaxation and aggregation of a simple polymer model. In this section, we briefly review the work by Ma and Hu [66, 67, 68, 69, 70] about relaxation and aggregation of polymer chains.

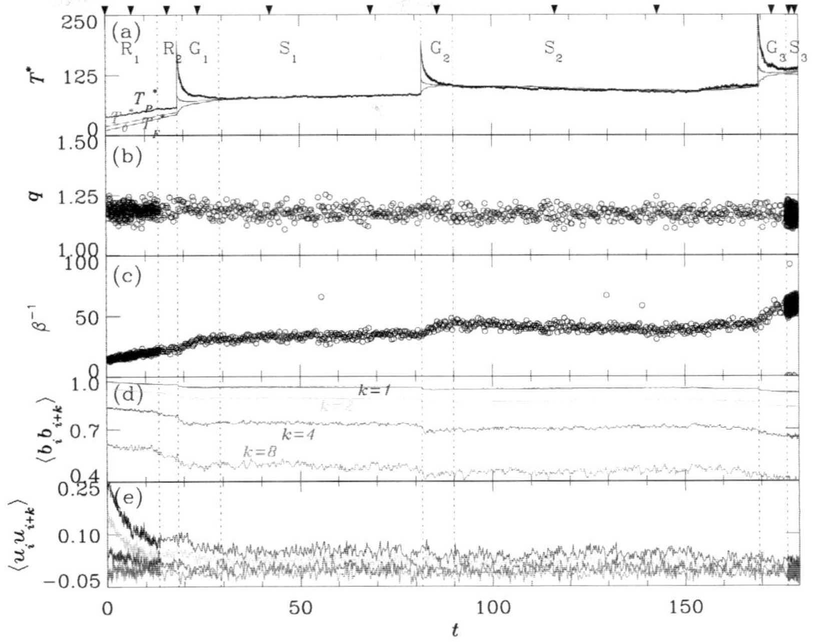

Figure 4. (Color online) Physical quantities versus time t in the unit of τ. (a) Temperature-like quantities T_0^* (red line), T_P^* (black line) and T_F^* (blue line) in unit ϵ for the whole system, for the polymer sites and for the fluid particles, respectively, in a single step; (b) q; (c) β^{-1}; (d) bond-pair correlation and (e) local helicity-pair correlation, for index distance $k = 1, 2, 4, 8$, in purple, green, blue and red colors, respectively. Data are plotted every 100 steps for (a), and 10000 steps for (b), (c), (d) and (e). The vertical dotted lines mark the change in integration time step δt from $(\delta t)_1 \equiv 5.0 \times 10^{-6}\tau$ to $(\delta t)_2 \equiv 2.5 \times 10^{-5}\tau$ at time $t = 13.4\tau$, dividing the relaxation stage into R_1 and R_2; the intermediate transients, G_1 (from $t = 18.4750$ to $t = 29.2925$), G_2 ($t = 81.6725$ to $t = 89.9700$) and G_3 ($t = 169.315$ to $t = 176.650$) that are followed by the quasi-static stages S_1, S_2 and S_3, respectively. In S_3, the integration step is reduced to $(\delta t)_3 \equiv 5 \times 10^{-7}\tau$. The velocity distributions at those steps marked by inverted triangles on top of plot (a) are analyzed in Fig. 5. Reproduced from Fig. 1 of [66].

3.1. Model System with monomers connected by rigid bonds

In this subsection, we consider that polymer molecules are chains of rigidly connected monomers, interacting via L-J potentials between non-bonded monomer pairs and bending and torsion potentials among consecutive sites along the chains. The polymers were modified from the united atom model [71] where the bond length and the interaction potentials describe polyethylene. The L-J potential [35, 36] between non-bonded monomer pairs at sites i and j with the distance r_{ij} is

$$v_{ij}(r_{ij}) = 4\epsilon((\frac{\sigma}{r_{ij}})^{12} - (\frac{\sigma}{r_{ij}})^6), \tag{1}$$

where σ and ϵ were used as units for length and energy strength, respectively. The bond length between two neighboring monomers in a polymer chain was fixed at $b_0 = 0.357\sigma$.

For consecutive sites i, $j = i + 1$, $k = i + 2$, and $l = i + 3$ along a polymer chain, the unit bond vector $\vec{b}_i$ defines the direction for two consecutive sites indexed by i and $j = i + 1$ along

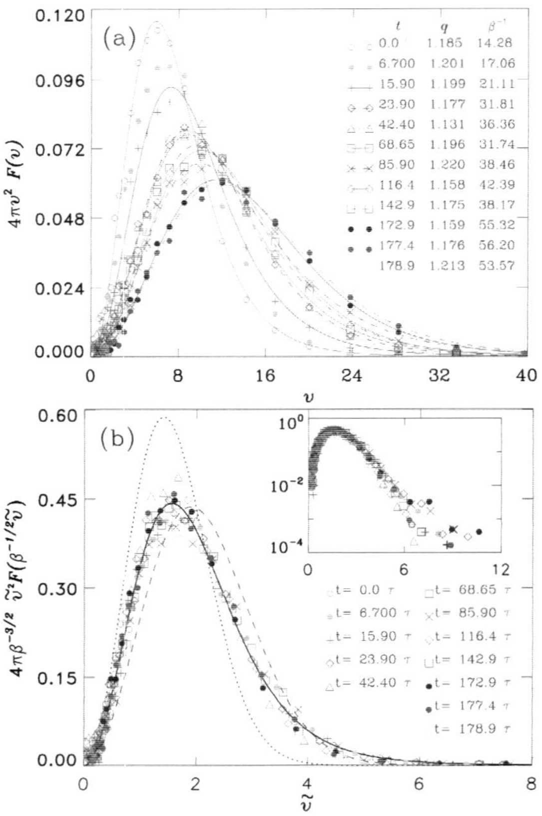

Figure 5. (Color online)(a) Velocity distributions and (b) scaled velocity distributions, for the system described in Fig. 4. The curves in (a) are the fitted functions (Eq. (4)) for the simulated data (in symbols) at the twelve time spots marked in Fig. 1(a). The corresponding time t (in τ) and fitted values q and β^{-1} (Fig. 1 (b) and (c)) for those curves are listed. The distributions in (b) are rescaled according to Eq. (6), where the solid line corresponds to $(q, \beta) = (1.1774, 1.0)$ and the dotted line represents the Maxwell-Boltzmann (MB) distribution with $(q, \beta) = (1.0, 1.0)$. We also plot the MB distribution with $(q, \beta) = (1.0, 0.554)$ as the dashed line, which has the same second moment $< \tilde{v}^2 >$ (or $\tilde{T}_P^*$) as that of solid line. The inset displays the simulation data in log scales. Note that the deviations from the main curves at time $t = 23.9$, 85.9 and 172.9 are contributed by the locally agitated events over the transient periods G_1, G_2 and G_3, respectively. Reproduced from Fig. 2 of [66].

the chain. The unit local helicity vector $\vec{u}_i$ is defined by

$$\vec{u}_i = \frac{\vec{b}_i \times \vec{b}_{i+1}}{|\vec{b}_i \times \vec{b}_{i+1}|}.$$

The bending and torsion angles are determined by

$$cos\theta_{ijk} = \vec{b}_i \cdot \vec{b}_j, \;\&\; cos\phi_{ijkl} = \vec{u}_i \cdot \vec{u}_j,$$

respectively. $180^o - \theta_{ijk}$ and ϕ_{ijkl} are, respectively, the valence angle and torsional angle considered in [72]. We set the bending potential

$$\Theta(\theta_{ijk}) = K_{\rm b}1097\epsilon(cos\theta_{ijk} - cos\theta_0)^2, \tag{2}$$

and the torsional potential

$$\Phi(\phi_{ijkl}) = K_{\rm t}\sum_{\iota=0}^{3} a_\iota(cos\phi_{ijkl})^\iota, \tag{3}$$

with $a_0 = 18.635\epsilon$, $a_1 = 38.16\epsilon$, $a_2 = 10.3\epsilon$ and $a_3 = -67.075\epsilon$, as in [71], θ_0 is either $\theta_0^{(I)} = 112.813^o$ or $\theta_0^{(II)} = 67.187^o$. We considered $K_{\rm b}$ and $K_{\rm t}$ and θ_0 as the tuning parameters related to the polymer properties at longer time and length scales. A cutoff of 2.3σ was used for pair L-J interactions.

The length σ_F and strength ϵ_F of the L-J interactions between fluid molecules were expressed in units of σ and ϵ, respectively. All fluid molecules and monomers were assumed to have the same mass m. The composite quantity $\tau = \sqrt{m\sigma^2/\epsilon}$ was used as time unit. The purely repulsive interaction potential between a monomer and a fluid molecule takes only the -12 power term in Eq. (1). In such homopolymer system, the thermodynamic conditions were changed by varying the volume (concentration) and the temperature. Heterogeneous conditions were introduced into the system by selecting a small portion of monomer sites, imposing different inter-site interactions. By mixing the polymers with spherical Lennard-Jones fluid molecules, we let the interaction potentials of fluid particles with the ordinary polymer sites be purely repulsive and with that small portions of specially selected sites ('linker sites') on polymers be the same as fluid-fluid potentials. Such setup mimics heterogeneities in the hydrophobic-hydrophilic effects in systems of biopolymers and help to stabilize the system. The shape and the strengths of all those interaction potentials are tunable as the means to alter the physical properties of the simulated system.

It is well known that the dynamic properties are related to the detailed numerical implements in solving the equations of motion in MD simulations [73]. In the simplified model, it should be emphasized that the well-tracked formulation of the discretized equations of motion that are treated as iterative dynamic systems. This aspect is appropriate in searching for microscopic origins of q-statistics [74]. The equations of motions in Lagrangian formulations were numerically solved using a velocity Verlet algorithm; the bonding between neighboring monomers was realized via the RATTLE scheme [75] to include the holonomic constraints for fixed bond lengths.

3.2. Relaxation of polymers with monomers in a chain connected by rigid bonds

We had chosen $K_{\rm b} = K_{\rm t} = 0.1$, corresponding to one tenth of those for polyethylene, so that the simulations can cover longer range of time scales. To start, we filled $N_P = 40$ polymer chains (each chain has $n = 100$ monomers) and $N_F/4$=1500 fluid molecules in a large enough cube of volume $V = L^3$ with periodic boundary conditions so that polymer chains can take some arbitrary random initial configurations. We took θ_0 in Eq. (2) to be $\theta_0^{(I)} = 112.813^o$ in all chains, selected $\sigma_F = \sigma/4$ and let a small N_L (=199) randomly selected 'linker sites' to have hydrophilic-like contacts with fluid particles.

We controlled the pressure and the kinetic energy [76] to adjust the volume V until it reached some value $V \approx (32.5\sigma)^3$. Then we replaced each fluid molecule with four closely packed molecules of zero kinetic energies and allowed the new system (of N_P polymers and N_F fluid molecules) to relax and reached finally a volume $V = L^3$ with $L = 33.3875\sigma$.

The NPT processes designed for equilibrium simulations [76] had been used in this preparation stage, where we controlled externally the pressure (P) and the temperature (T) to drive the system of fixed number of particles (N) to the target density. The process also fine-tuned the conformations of the polymer chains that helped the effective performance of RATTLE-algorithm in the subsequent NVE simulations.

We then started the production simulations with the integration time step $(\delta t)_1 \equiv 5.0\times10^{-6}\tau$ and focus on the NVE relaxation processes with constant volume V and total energy E without any external controlling. To speed up the simulation process, we increased the integration time step δt from $(\delta t)_1 \equiv 5.0\times10^{-6}\tau$ by five times to $(\delta t)_2 \equiv 2.5\times10^{-5}\tau$ at time $t = 13.4\tau$.

The production simulations from $t = 0$ continued for about 180 τ (at the final few τ the integration time step was reduced, see below). The results were shown in Figs. 1 and 2 of [66], reproduced as Figs. 4 and 5 in the present paper. Figure 4(a) shows that the time evolution of the temperature-like quantities, T_P^*, T_F^* and T_0^*, consist of several stages, where T_P^* , T_F^*, and T_0^* are twice of the mean kinetic energies per degrees of freedom for polymer chains, fluid molecules, and the whole system, respectively. For polymer chains, $T_P^* \equiv f^{-1}\sum_{\text{site } i} m_i v_i^2$, where $f = N_P(2n+1)$ is total number of degrees of freedom for polymer chains when $(n-1)$ holonomic constraints for each polymer chain were taken into account.

In the first stage R_1, the local conformation relaxes with decreasing correlations in bond-pairs (Fig. 4(d)) and local helicity pairs (Fig. 4(e)). We used inner product $\vec{b}_i\cdot\vec{b}_{i+k}$ and $\vec{u}_i\cdot\vec{u}_{i+k}$ for indices i and $i+k$ along a chain to define the correlations of index distance k.

The increase of the integration time step at $t = 13.4\tau$ would certainly affect the route of time evolution at microscopic level. Our focus here is whether the value of q changes. The subsequent relaxation stage R_2 continues before the system becomes quasi-static with $T_P^* \approx T_F^*$. The system is found subject to intermediate transients before and after reaching a quasi-static stage; there are three quasi-static stages, S_1, S_2 and S_3, following the transient stages G_1, G_2 and G_3, respectively. Entering the static stage S_3, we reduced the integration time step by a factor fifty to $(\delta t)_3 \equiv 5.0\times10^{-7}\tau$, to understand its effect on q. From simulation data, we estimated the diffusion constant for monomers as $D \approx 3.3\sigma^2/\tau$. The time spans of S_1 and S_2 are large enough for a monomer to cross a distance ranging from $L/2$ to L.

The vertical dotted lines in Fig. 4 mark the change in integration time step δt from $(\delta t)_1 \equiv 5.0\times10^{-6}\tau$ to $(\delta t)_2 \equiv 2.5\times10^{-5}\tau$ at time $t = 13.4\tau$, dividing the relaxation stage into R_1 and R_2; the intermediate transients, G_1 (from $t = 18.4750$ to $t = 29.2925$), G_2 ($t = 81.6725$ to $t = 89.9700$) and G_3 ($t = 169.315$ to $t = 176.650$) that are followed by the quasi-static stages S_1, S_2 and S_3, respectively. The velocity distributions at those steps marked by inverted triangles on top of Fig. 4(a) were analyzed in Fig. 5.

Figure 5(a) shows that the velocity distributions of monomers at several typical time steps (marked by inverted triangles at the top line of Fig. 4) can be well described by Tsallis q-statistics [77] with a generalized Maxwell-Boltzmann distribution (GMBD) [78, 79, 80]

$$4\pi v^2 F(v) = A_q \beta^{3/2} 4\pi v^2 \times [e_q^{-\frac{\beta}{2}mv^2}] \tag{4}$$

where $e_q^x \equiv [1+(1-q)x]^{\frac{1}{(1-q)}}$, $A_q = [\int_0^\infty 4\pi v^2 e_q^{-\frac{1}{2}mv^2} dv]^{-1}$, and $1 \le q < 7/5$ [66]. In the limit $q \to 1$, Eq. (4) becomes the Maxwell-Boltzmann (MB) distribution

$$4\pi v^2 [(\frac{m\beta}{2\pi})^{\frac{3}{2}} \times e^{-\frac{\beta}{2}mv^2}]. \tag{5}$$

The parameter β in Eq. (4) can be absorbed to scale the velocity variable v,

$$\beta^{-\frac{1}{2}} 4\pi v^2 F(v) = A_q 4\pi \tilde{v}^2 e_q^{-\frac{\tilde{v}^2}{2}}, \tag{6}$$

where $\tilde{v} = \beta^{\frac{1}{2}} v$ and the right-hand side (RHS) of Eq. (6) is free of β, *i.e.*, a family of distributions with different β but same q can collapse into one master curve.

Figure 5(b) shows that the velocity distributions in Fig. 5(a) collapse into a single curve, after being scaled by using Eq. (6). The parameter β^{-1} increases linearly with the instantaneous

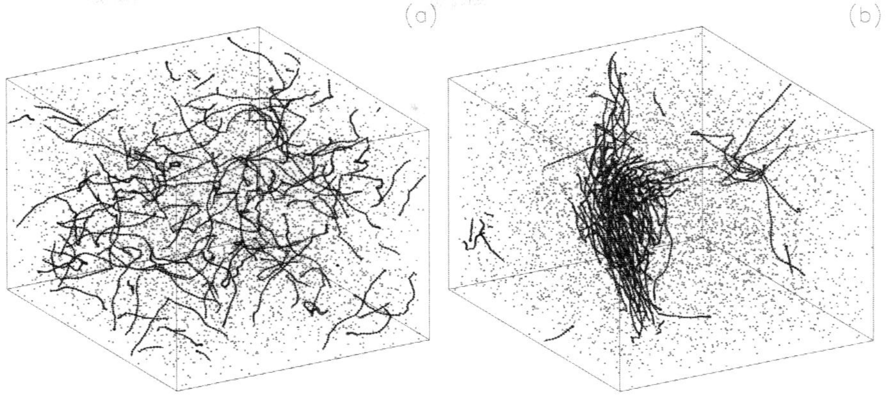

Figure 6. Snapshots of the configurations before (a) and after (b) the formation of polymer aggregation, during the phase separation process for a system with N_P =40 chains, each with n =100 rigidly bonded monomers, mixed with a fluid of N_F=6000 spherical atoms (shown as grey dots). The monomers along a chain are also subject to bending and torsion (intra-chain) potentials in very weak strengths. Periodic boundary condition is applied to each of the three directions for the $33.3876\sigma \times 33.3876\sigma \times 33.3876\sigma$ cell. The instantaneous temperatures for polymer and fluid are $T_P^* = 18.21$ and $T_F^* = 19.95$ for (a); and $T_P^* = 6.335$ and $T_F^* = 8.24$ for (b), respectively. Reproduced from Fig. 1 of [68].

temperature of monomers T_P^* [66]. The result in Fig. 5(b) suggests that Eq. (4) should not be considered merely as a class of fitting functions, it relates the microscopic states over a wide time span as a single non equilibrium state characterized by the family of distributions of virtually the same q. Indeed, the reducing of integration time step δt can change the heights and locations of regions G_1, G_2 and G_3 in Fig. 4(a), but does not change q, see e.g. S_3 in Fig. 4(a) and 4(b). As for the physical origin of such GMBD, it is useful to consider the thought of Beck and Cohen [81].

Note that each distribution in Fig. 5 is over the monomers in the system for one single step. If it is also a probability distribution for an individual monomer, then the x-component of the velocity of particle i satisfies

$$< v_{ix}{}^2 >= C_{q,\beta} \int v_{ix}{}^2 e_q^{-\beta K} \prod_{j=1}^{N} dv_{jx} dv_{jy} dv_{jz},$$

where K is the total kinetic energy of the polymers. Using $(e_q^x)' = e_q^{qx}$ or $(e_q^x)' = \frac{e_q^x}{1+(1-q)x}$ and integration by parts, we derived that the mean kinetic energy of a monomer is

$$< \frac{1}{2} m v^2 >= \frac{3}{2(2-q)} [\frac{1}{\beta} - (1-q) < K >]. \qquad (7)$$

Equation (7) states a modified version of equipartition theorem; it becomes the conventional equipartition theorem only when $q = 1$. The conditions for the validity of Eq. (7), especially the role of heterogeneity [82], deserves further analysis.

3.3. Aggregation of polymer chains connected by rigid-bonds

We take a system at the end of simulation of Fig. 4 ($t = 180\tau$) as the initial system and reduce K_b and K_t to 1.0×10^{-4}. With quenching and heating alternatively, each for about 50 τ, we let

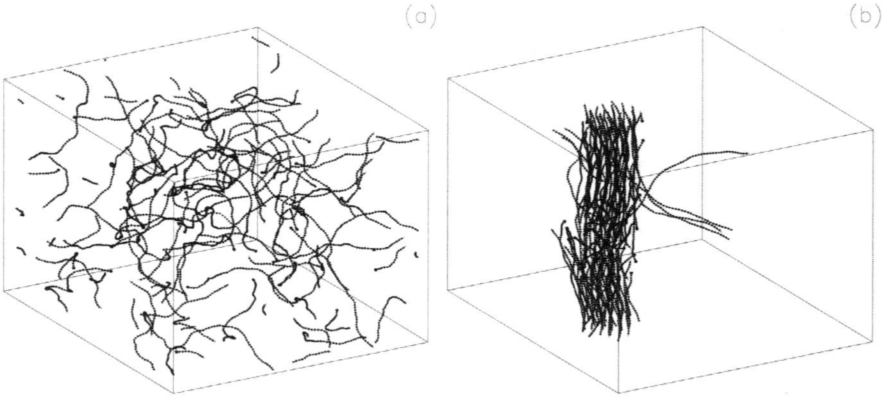

Figure 7. Snapshots before (a) and after (b) the formation of aggregation, for a system similar to that in Fig. 6, except the fluid atoms are absent and the polymer sites are identical and subject to neither bending nor torsion potentials. The instantaneous temperature is $T_P^* = 26.78$ for (a) and $T_P^* = 6.78$ for (b), respectively. Reproduced from Fig. 2 of [68].

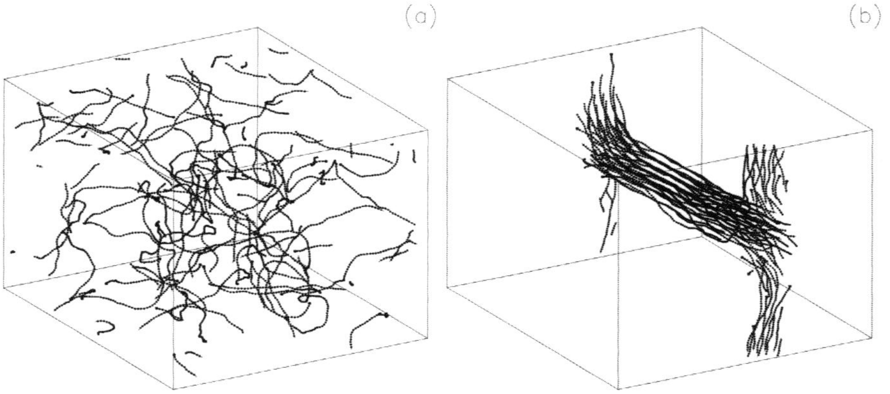

Figure 8. Snapshots before (a) and after (b) the formation of aggregation, for a system similar to that in Fig. 7, but the polymer chains are subject to the same bending and torsion angle potentials as those in the mixed system plotted in Fig. 6. The polymer temperature for (a) is $T_P^* = 21.7$ and for (b) is $T_P^* = 6.42$. Reproduced from Fig. 3 of [68].

the system free from possible residual local structures favored by the original angle potentials. We then let the system ready to relax. We also consider the case of pure polymers with zero (K_b = K_t=0) or very small ($K_\mathrm{b} = K_\mathrm{t} = 10^{-4}$) angle potentials in absence of fluid. It is found that, in all cases, very small or zero angle potentials render the systems to undergo slow quenching processes, as a dynamical system driven by a low temperature attractor caused by the numerical sink. Such scenario is also true for other cases considered in this subsection, as soon as the strengths of angle potentials are tiny.

Since integration of time step δt is relevant in determining the time evolution of the aggregation process, we will consistently use two major integration time steps $(\delta t)_2 = 2.5 \times 10^{-5}$ and $(\delta t)_4 = 1 \times 10^{-4}$. Over the beginning a change (on the strengths of angle potential, or on the lengths of polymer chains, etc...) being made, we use the smaller one $(\delta t)_2$ and turn to the large one afterward. While the quenching rate depends on the choice of δt, the features displayed on the aggregated states would not.

We examine the snapshots of the three cases: the mixed system of polymers and fluids (Fig. 6), the pure system where the monomers at linker sites are switched into ordinary monomers and $K_{\rm b}$ and $K_{\rm t}$ are set to 0 (Fig. 7), and the pure system where the monomers at linker sites are switched into ordinary monomers and $K_{\rm b}$ and $K_{\rm t}$ are kept at 10^{-4} (Fig. 8) [68]. In all of three cases, polymer chains aggregate as K_b and K_t become 0 or very small.

3.4. Polymer chains with monomers connected by springs

To check whether the obtain q-statistics with $q > 1$ is an artifact due to the RATTLE scheme [75] or rigid-bond constraints between neighboring monomers, we used the method of ordinary MD simulations to study the relaxation process in a system of polymer chains and Lennard-Jones molecules, in which each polymer chain consists of n monomers labelled by integer $i = 1, \ldots, n$ from one end to the other; the monomers at the sites i and $j = i + 1$ $(1 \leq i < n)$ are connected by harmonic potentials

$$\frac{1}{2} k_{spring}(r_{ij} - r_0)^2. \tag{8}$$

The coupling constant k_{spring} is $k_s = 10^s \times k_0$, with $s = 0, 1, 2, 3, 4$ for the five systems, respectively. Please note that the interactions between neighboring monomers of the form of Eq. (8) has been used in the Go-like model for proteins [83] and also in a RNA model [84]. L-J interactions, and bending-angle and torsional-angle dependent interactions are the same as those described in previous subsections.

The instantaneous temperatures of the monomers in polymer chains, T_P^*, and the fluid, T_F^*, were initially different, but they vary in time and their ratio $\Gamma^* = T_P^*/T_F^*$ approached to a constant during the relaxation process. The velocity distributions of monomers in the relaxation process follow q-statistics [77] with $q \geq 1$. The value of q and the limiting ratio of Γ^* depend on k_{spring}; in the strong strength limit, they approach those for the system, in which each polymer chain consists of monomers connected by rigid bonds; in the weak strength limit, Γ^* and q approach 1 corresponding to MB distribution for velocities of monomers. Typical simulation results are shown in Figs. 5, 6 and 7 of [67]. Such results indicate that there is a continuous increase of q as one increases the strength of springs (between monomers) from the smaller to the larger values, and the result $q > 1$ in previous subsections is not an artifact of the RATTLE scheme [75] or the rigid-bond constraints.

It was found that when neighboring monomers of the spring-connected polymer chains have small or zero bending-angle and torsion-angle dependent potentials, then the polymer chains also aggregate [69]. When $K_b = K_t = 0.1$, polymer chains do not aggregate even at low temperatures. Thus K_b and K_t play important role for the aggregation of polymer chains.

In summary, the molecular dynamics was used to simulate various systems of polymer chains and L-J molecules; the neighboring monomers along a polymer chain are connected by rigid bonds [66] or spring of strength k_{spring} [67]. It was found that during the relaxation process the velocity distributions of monomers in a wide range of simulation time can be well described by Tsallis q-statistics [77] with a generalized Maxwell-Boltzmann distribution (GMBD) [78, 79, 80] and a single scaling function. The value of q is related to the conformation constraining potential, the interactions with background fluid, the destruction of chain homogeneity or the value of k_{spring}. Such study is useful for understanding microscopic origin of q-statistics [81, 74, 85]. The results suggest that complex non-equilibrium systems can still have simple scaling behavior and inspire many interesting problems for further studies.

The system studied in [66] has slow relaxation, similar to that of a spin-glass model at low temperatures [86]. The glassy behavior has also been found in the collagen fibril [87, 88]. It seems

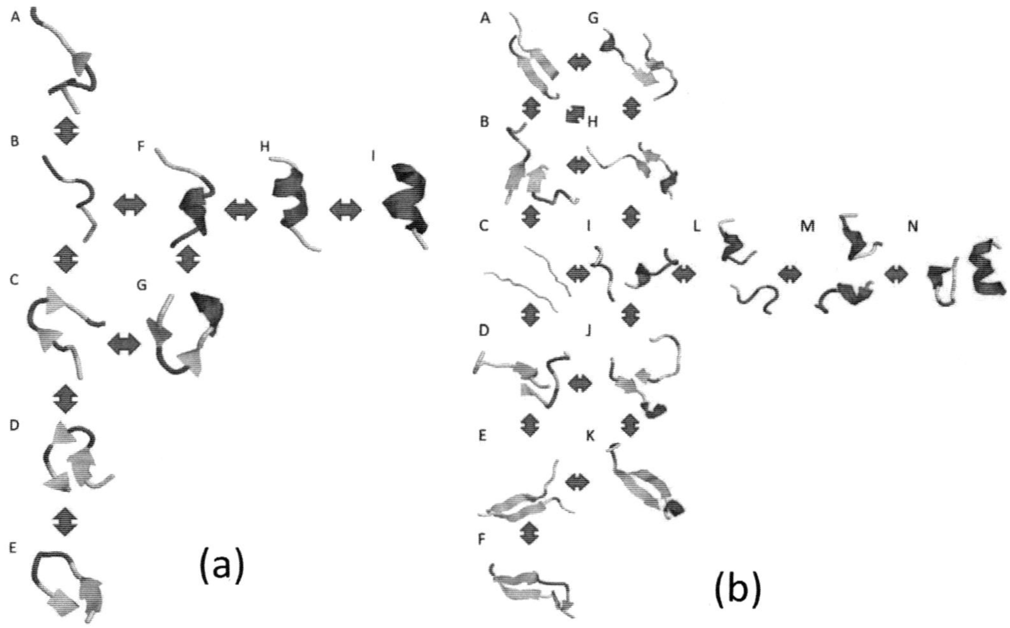

Figure 9. (Color online) (a) Conformations and the transformation pathways of the global- and local-minimum free-energy states A to I of the the system one polyQ-peptide and 1611 water molecules. (b) Conformations and the transformation pathways of the global- and local-minimum free-energy states A to N of the system with two polyQ-peptides and 3129 water molecules. The peptide conformations were created using RasMol [91]. (a) and (b) were reproduced from Fig. 2 and Fig. 4 of [15].

that the glassy behavior of bio-polymers is the key factor for a biological system to maintain in a non-equilibrium state [23, 51].

It was also found that when neighboring monomers of a polymer chain have small or zero bending-angle and torsion-angle dependent potentials, then the polymer chains tend to aggregate [68, 69, 70]. The aggregation of proteins in human brains can result in neurodegenerative diseases, such as Alzheimer's disease [89, 6, 7], Huntington's disease [8], Parkinson's disease [16, 17], etc. Thus it is of interest to know relaxation and aggregation behaviors of polymer chains in non-equilibrium situation.

4. All atom models for protein aggregation

In this section, we briefly review recent developments for simulations of all atom models for protein aggregation.

structure	probability(%)	ΔF(kcal/mol)
helix	29.4±4.6	0.47±0.12
intra-β-sheet	6.3±5.1	1.39±0.49
random-coil	64.9±8.9	-

Table 1. Probability and free energy difference of each secondary structure in the one polyQ-peptide system.

4.1. Oligomerization of peptides LVEALYL and RGFFYT and their binding affinity to insulin. Recently it has been proposed a model for fibrils of human insulin in which the fibril growth proceeds via stacking LVEALYL (fragment 11-17 from chain B of insulin) into pairs of tightly interdigitated b-sheets. The experiments have also shown that LVEALYL has high propensity to self-assembly and binding to insulin. This necessitates study of oligomerization of LVEALYL and its binding affinity to full-length insulin. Using the all-atom simulations with Gromos96 43a1 force field and explicit water it is shown that LVEALYL can aggregate. Theoretical estimation of the binding free energy of LVEALYL to insulin by the molecular mechanic Poisson-Boltzmann surface area method reveals its strong binding affinity to chain B, implying that, in agreement with the experiments, LVEALYL can affect insulin aggregation via binding mechanism. We predict that, similar to LVEALYL, peptide RGFFYT (fragment B22-27) can self-assemble and bind to insulin modulating its fibril growth process. The binding affinity of RGFFYT is shown to be comparable with that of LVEALYL [90].

4.2. MD simulations of one and two polyglutamine peptides in explicit water molecules
Aggregation of polyglutamine peptides with β-sheet structures is related to some important neurodegenerative diseases such as Huntingtons disease. However, it is not clear how polyglutamine peptides form the β-sheets and aggregate. To understand this problem, we performed all-atom replica-exchange molecular dynamics simulations of one and two polyglutamine peptides with 10 glutamine residues in explicit water molecules. Conformations and the transformation pathways of the global- and local-minimum free-energy states A to I of the the system with one polyQ-peptide and 1611 water molecules was shown in Fig. 9(a); Conformations and the transformation pathways of the global- and local-minimum free-energy states A to N of the system with two polyQ-peptides and 3129 water molecules was shown in Fig. 9(b).

Our results show that two polyglutamine peptides mainly formed helix or coil structures when they are separated, as in the system with one-polyglutamine peptide. As the interpeptide distance decreases, the intrapeptide β-sheet structure sometimes appear as an intermediate state, and finally the interpeptide β-sheets are formed. We also find that the polyglutamine dimer tends to form the antiparallel β-sheet conformations rather than the parallel β-sheet, which is consistent with previous experiments and a coarse-grained molecular dynamics simulation [15].

Table 1 shows that in the one polyQ-peptide system, the random-coil has the highest probability of 64.9±8.9, the helix has the second probability of 29.4±4.6, and intra-β-sheet has probability of only 6.3±5.1 Table 2 shows that in the two polyQ-peptide system, inter-β-sheet has the highest probability of 50.5±10.2, random-coil has the probability of 25.3±10.7, the helix has the probability of 23.1±3.2, and intra-β-sheet has the probability of 11.8±5.7. Thus the most stable conformation of two polyQ-peptide system, i.e. inter-β-sheet, is different from the most stable conformation of one polyQ-peptide system, i.e. the random-coil.

structure	probability(%)	ΔF(kcal/mol)
helix	23.1±3.2	0.47±0.15
inter-β-sheet	50.5±10.2	-
intra-β-sheet	11.8±5.7	0.87±0.31
random-coil	25.3±10.7	0.41±0.28

Table 2. Probability and free energy difference of each secondary structure in the two polyQ-peptide system.

4.3. MD approach to effect of Taiwan mutation (D7H) on structures of amyloid peptides
Recent experiments have shown that the Taiwan mutation (D7H) slows the fibril formation of amyloid peptides Aβ40 and Aβ42 [92]. Motivated by this finding, we have studied the influence of D7H mutation on structures of Aβ peptide monomers using the replica exchange molecular dynamics simulations with OPLS force field and implicit water model. Our study reveals that the mechanism behind modulation of aggregation rates is associated with decrease of β-content and dynamics of the salt bridge D23-K28. Estimating the bending free energy of this salt bridge, we have found that, in agreement with the experiments, the fibril formation rate of both peptides Aβ40 and Aβ42 is reduced about two times by mutation [93].

5. Searching potential drugs for AD

From database of about 32364 natural compounds [94], the Lipinskis rule of five [95] was applied to reduce the number of compounds from 32364 to 3699 which was further studied by the docking method.

In order to study the binding affinity to mature fibrils, two typical structures of 6Aβ_{9-40} obtained from Prof. R. Tycko [96] and 5Aβ_{17-42} (PDB ID: 2BEG [97]) were used.

A ranking of ligands by their binding energies to two receptors [96, 97] was made and the 10 top leads were listed on Table S1 of the Supporting Information (SI) in [98]. Among such 10 top leads only Dihydrochalcone could be purchased in the market.

.

The binding affinity of this ligand (Dihydrochalcone) to amyloid beta (Aβ) fibril has been thoroughly studied by computer simulation and experiment. Using the Thioflavin T (ThT) assay we had obtained the inhibition constant IC50$\approx 2.46\mu M$. This result is in good agreement with the estimation of the binding free energy obtained by the molecular mechanic-Poisson Boltzmann surface area method and all-atom simulation with the force field CHARMM 27 and water model TIP3P. Cell viability assays indicated that Dihydrochalcone could effectively reduce the cytotoxicity induced by Aβ. Thus, both in silico and in vitro studies show that Dihydrochalcone is a potential drug for the Alzheimers disease [98].

Dihydrochalcone is a compound from Chinese medicine *Xuejie* (English Dragon's Blood; Latin Draconis Sanguis). According to Chinese historical record, *Xuejie* was already used as a medicine in Ming Dynasty. Dihydrochalcone is also the active ingredient of a traditional herbal tea in Africa.

6. Summary and discussion

It is pointed out in Section 1 that universality and scaling are key concepts developed from the study of phase transitions and critical phenomena in physical critical systems. Using computing packages developed by us [99], we have found a universal volume to surface area for proteins in Protein Data Bank [100]. We have used simple lattice models and polymer chain model to understand some key factors for protein aggregation. In the next step, it is of interest to develop universal theory for protein aggregation.

Figures 4 and 5 show that velocity distributions of monomers in polymer chains deviate from Maxwell-Boltzmann distribution. Since the data of Figures 4 and 5 were obtained by molecular dynamics simulation and the simulated systems still do not reach equilibrium. It is of interest to use other simulation methods, such as parallel tempering which can allow the simulated systems to reach equilibrium more quickly, to study the same polymer systems to check whether in such cases the velocity distributions of monomers in polymer chains still deviate from Maxwell-Boltzmann distribution.

It had been proposed that protein aggregation is due to protein misfolding [101, 102]. However, Figures 1(a) and 2(a) show that peptide chains aggregate with the conformation of the first excited state of the one chain system rather than the ground state conformation. Figure 9, Table 1, and Table 2 show that most stable conformation of one chain system is different from the most stable conformation of two-chains system. In such cases, protein aggregation is not due to protein mis-folding.

We have found Dihydrochalcone as a potential drug for AD [98] from *in silico* and *in vitro* study. It is of interest to use animal model to study the effect of Dihydrochalcone to prevent AD. Table S1 in SI of [98] list other 9 compounds besides Dihydrochalcone. These compounds are also potential drugs for AD. It is of interest to study the effect of such compounds to reduce protein aggregation via *in vitro* and *in vivo* study.

7. Acknowledgments

The author would like to thank collaborators, M. S. Li, N. T. Co, G. Reddy, J. E. Straub, D. Thirumalai, W.-J. Ma, H. -L. Chiang, S. T. Ngo, C. -J. Chen, M. H. Viet, C.-Y. Chen, Y.-R. Chen, H. Okumura, P. M. Truong, M. H. Viet, P. H. Nguyenand, et al. for the work reported in this brief review paper. The author would like to thank Wei Li for inviting him to give a talk at The 4th International Workshop on Statistical Physics and Mathematics for Complex Systems (SPMCS2014), in Yichang, China on 13-16 October 2014; this paper was based on the talk presented at SPMCS2014. This work was supported by Grant MOST 103-2112-M-001 -016.

References

[1] Alzheimer A 1907 *Allg. Z. Psychiatr.* **64** 146
[2] Alzheimer A 1911 *Z. Ges. Neurol. Psychiat.* **4** 356
[3] Kraepelin E 1910 *Psychiatrie. Ein Lehrbuch fur Studierende und AL rzte. II. Band* (Barth Verlag, Leipzig)
[4] Goedert M and Grazia M 2006 *Science* **314** 777-781
[5] Roberson ED and Mucke L 2006 *Science* **314** 781-784
[6] Petkova AT, *et al.* 2002 *Proc. Natl. Acad. Sci. 99* 16742
[7] Luhrs T, *et al.* 2005 *Proc. Natl. Acad. Sci. 102* 17342
[8] Huntington G 1872 *Medical and Surgical Reporter of Philadelphia* **26** 317-321
[9] MacDonald ME, Gusella JF 1996 *Curr. Opin. Neurobiol.* **6** 638
[10] Koshy BT, Zoghbi HY 1997 *Brain Pathol.* **7** 927
[11] Kakizuka A 1998 *Trends Genet.* **14** 396
[12] Perutz MF 1999 *Trends Biochem. Sci.* **24** 58
[13] Paulson HL 1999 *Am. J. Hum. Genet.* **64** 339
[14] Riley BE and Orr HT 2006 *Genes Dev.* **20** 2183
[15] Chiang HL, Chen CJ, Okumura H and Hu CK 2014 *J. Comp. Chem.* **35** 1430-1437
[16] Parkinson J 1817 *An Essay on the Shaking Palsy* (Sherwood, Neely, and Jones, London)
[17] Polymeropoulos MH, *et al.* 1997 *Science* **276** 2045-2047
[18] Irvine GB, El-Agnaf OM, Shankar GM and Walsh DM 2008 *Mol. Med.* **14** 451-464
[19] Leslie M 2008 *Science* **321** 1764-1765
[20] Boltzmann L (1896, 1898) *Lectures on gas theory* translated by Stephen G. Brush (1964) Berkeley: University of California Press; (1995) New York: Dover
[21] Gibbs JW 1902 *Elementary principles in statistical mechanics* (New York), reprinted (Woodbridge, CT, Ox Bow Press, 1981)
[22] Stanley HE 1971 *Introduction to Phase Transitions and Critical Phenomena* (Oxford Univ. Press, New York)
[23] Hu CK 2014 *Chinese J. Phys.* **52** 1-76

[24] Wilson KG 1971 *Phys. Rev. B* **4** 3174 and 3184
[25] Wilson KG and Fisher ME 1972 *Phys. Rev. Lett.* **28** 240
[26] Wilson KG 1972 Phys. Rev. Lett. **28** 548
[27] Wilson KG 1983 *Rev. Mod. Phys.* **55** 583
[28] Guggenheim EA 1945 *J. Chem. Phys.* **13** 253
[29] Yang CN and Lee TD 1952 *Phys. Rev.* **87** 404; Lee TD and Yang CN 1952 *Phys. Rev.* **87** 410
[30] Ising E 1925 *Zeitschrift fur Physik* **31** 253
[31] Blöte HWJ, Luijten E and Heringa JR 1995 *J. Phys. A: Math. Gen.* **28** 6289-6313
[32] Talapov AL and Blöte HWJ 1996 *J. Phys. A: Math. Gen.* **29** 5727
[33] Sengers JV and Shanks JG 2009 *J. Stat. Phys.* **137** 857
[34] Watanabe H, Ito N, and Hu CK 2012 *J. Chem. Phys.* **136** 204102
[35] Jones JE 1924 *Proc. R. Soc. London A* **106** 463
[36] Lennard-Jones JE 1931 *The Proceedings of the Physical Society* **43** part 5
[37] Metropolis N and Ulam S 1949 *Journal of American Statistical Association* **44** 335
[38] Metropolis N, Rosenbluth AW, Rosenbluth MN, Teller AH and Teller E 1953 *J. Chem. Phys.* **21** 1087
[39] Hu CK 1992 *Phys. Rev. B* **46** 6592-6595
[40] Hu CK 1992 *Phys. Rev. Lett.* **69** 2739-2742
[41] Wu MC and Hu CK 2002 *J. Phys. A.* **35** 5189-5206
[42] Ivashkevich EV, Izmailian NS and Hu CK 2002 *J. Phys. A.* **35** 5543-5561
[43] Hu CK 1994 *J. Phys. A* **27** L813-L820
[44] Hu CK, Lin CY and Chen JA 1995 *Phys. Rev. Lett.* **75** 193-196
[45] Hu CK and Lin CY 1996 *Phys. Rev. Lett.* **77** 8-11
[46] Wang FG and Hu CK 1997 *Phys. Rev. E* **56** 2310-2313
[47] Okabe Y, Kaneda K, Kikuchi M and Hu CK 1999 *Phys. Rev. E* **59** 1585-1588
[48] Tomita Y, Okabe Y and Hu CK 1999 *Phys. Rev. E* **60** 2716-2720
[49] Izmailian NS and Hu CK 2001 *Phys. Rev. Lett.* **86** 5160-5163
[50] Wu MC, Hu CK and Izmailian NS 2003 *Phys. Rev. E* **67** 065103
[51] Hu CK 2013 *AIP Conf. Proc.* **1518** 541
[52] Flory PJ 1949 *J. Chem. Phys.* **17** 1347-1348
[53] Flory PJ and Krigbaum WR 1949 *J. Chem. Phys.* **18** 1086-1094
[54] Shakhnovich E and Gutin A 1990 *J. Chem. Phys.* **93** 5967
[55] Vogel T, Bachmann M and Janke W 2007 *Phys. Rev. E* **76** 061803
[56] Lee JH, Kim SY and Lee J 2010 *J. Chem. Phys.* **133** 114106
[57] Lee JH, Kim SY and Lee J 2012 *Phys. Rev. E* **86** 011802
[58] Bedini A, Owczarek AL and Prellberg T 2013 *J. Phys. A: Math. Gen.* **46** 085001
[59] Chen CN, Hsieh YH and Hu CK 2013 *EPL* **104** 20005
[60] Li MS, Co NT, Reddy G, Hu CK, Straub JE and Thirumalai D 2010 *Phys. Rev. Lett.* **105** 218101
[61] Chiti F, Stefani M, Taddei N, Ramponi G and Dobson CM (2003) *Nature* **424** 805
[62] Co NT, Hu CK and Li MS 2013 *J. Chem. Phys.* **138** 185101
[63] Cabaleiro-Lago C, Quinlan-Pluck F, Lynch I, Dawson KA and Linse S 2010 *ACS Chem. Neurosci.* **1** 279-287
[64] Sallon S, *et al.* (2008) *Science* **320** 1464
[65] Yashina S, *et al.* 2012 *Proc. Nat. Acad. SCi. USA* **109** 4008-4013
[66] Ma WJ and Hu CK 2010 *J. Phys. Soc. Japan* **79** 024005
[67] Ma WJ and Hu CK 2010 *J. Phys. Soc. Japan* **79** 024006
[68] Ma WJ and Hu CK 2010 *J. Phys. Soc. Japan* **79** 054001
[69] Ma WJ and Hu CK 2010 *J. Phys. Soc. Japan* **79** 104002
[70] Hu CK and Ma WJ 2010 *Prog. Theor. Phys. Supp.* **184** 369
[71] Steele D 1985 *Chem. Soc. Faraday Trans. 2* **81** 1077
[72] Karatasos K, Adolf AB and Hotston S 2000 *J. Chem. Phys.* **112** 8695
[73] Schlick T 2002 *Molecular Modelling and Simulation: an Interdisciplinary Guide* (Springer, New York)
[74] Mayoral E and Robledo A 2005 *Phys. Rev. E* **72** 26209; Grassberger P 2005 *Phys. Rev. Lett.* **95** 140601 (2005); Robledo A 2006 *Physica A* **370** 449
[75] H. C Andersen., J. Comput. Phys. **52**, 24 (1983).
[76] M. P. Allen and D. J. Tildesley, *Computer Simulation of Liquids* (Oxford Univ. Press, Oxford, 1987).
[77] Tsallis C 1988 *J. Stat. Phys.* **52** 479
[78] Lima JAS and Silva R 2005 *Physica A* **338** 272
[79] Silva R Jr, Plastino AR and Lima JAS 1998 *Phys. Lett. A* **249** 401
[80] Douglas P, Bergamini S and Renzoni F 2006 *Phys. Rev. Lett.* **96** 110601
[81] Beck C and Cohen EGD 2003 *Physica A* **322** 267

[82] Vignat C and Plastino A 2005 *Phys. Lett. A* **343** 411
[83] Go N 1983 *Ann. Rev. Biophys. Bioeng.* **12** 183
Kouza M, *et al.* 2005 *Biophysical J.* **89** 3353
Li MS, Hu CK, Klimov DK and Thirumalai D 2006 *Proc. Natl. Acad. Sci. U.S.A.* **103** 93
Li MS, Kouza M and Hu CK 2007 *Biophysical J.* **92** 547
Kouza M, Hu CK and Li MS 2008 *J. Chem. Phys.* **128** 045103
Fang HJ, *et al.* 2009 *Biophysical J.* **96** 1892
[84] Mamasakhlisov YS, Hayryan S, Morozov VF and Hu CK 2007 *Phys. Rev. E* **75** 061907
[85] Pluchino A, Rapisarda A and Tsallis C 2007 *EPL* **80** 26002
[86] Dasgupta C, Ma SK and Hu CK 1979 *Phys. Rev. B* **20** 3837
[87] Gevorkian SG, Allahverdyan AE, Gevorgyan DS and Hu CK 2011 *EPL* **95** 23001
[88] Gevorkian SG, Allahverdyan AE, Gevorgyan DS, Simonian AL and Hu CK 2013 *Plos One* **8** e78526
[89] Goate A, *et al.* 1991 *Nature* **349** 704
[90] Chiang HL, Ngo ST, Chen CJ, Hu CK and M. S. Li 2013 *Plos One* **8** 65358
[91] Sayle RA and Milner-White EJ 1995 *Trends Biochem. Sci.* **20** 374
[92] Chen WT *et al.* 2012 *PLoS One* **7** e35807
[93] Truong PM, Viet MH, Nguyenand PH, Hu CK and Li MS 2014 *J. Phys. Chem. B* **118** 8972
[94] Chen CYC 2011 *PLoS ONE* **6** e15939
[95] Lipinski A, Lombardo F, Dominy BW and Feeney PJ 1997 *Adv Drug Del Rev* **23** 3-25
[96] Petkova AT, Yau WM and Tycko R 2006 *Biochemistry* **45** 498-512
[97] Luhrs T, Ritter C, Adrian M, Riek-Loher D, Bohrmann B, et al. 2005 *Proc. Natl. Acad. Sci. U.S.A.* **102** 17342-17347
[98] Viet MH, Chen CY, Hu CK, Chen YR and Li MS 2013 *Plos One* **8** 79151
[99] Hayryan S, Hu CK, Skrivanek J, Hayryan E and Pokorny I 2005 *J. Compu. Chem.* **26** 334-343
Busa J, *et al.* 2005 *Compu. Phys. Commu.* **165** 59-96
Busa JJ, Hayryan S, Wu MC, Busa J and Hu CK 2012 *Compu. Phys. Commu.* **183** 2494-2497
[100] Wu MC, Li MS, Ma WJ, Kouza M and Hu CK 2011 *EPL* **96** 68005
[101] Dobson CM 2003 *Nature* **426** 884
[102] Chiti F and Dobson CM 2006 *Annu. Rev. Biochem.* **75** 333

A simple phenomenological model for time evolution of social networks

J. Jiang[1], Q. A. Wang[2] , W. Li[3] and X. Cai[3]

[1]Research Center of Nonlinear Science and College of Mathematics and Computer Science of Wuhan Textile University, Wuhan, 430200, China

[2]Institut Supérieur des Matériaux et Mécaniques Avancés 44, Avenue F.A. Bartholdi, 72000 Le Mans, France

[3]Complexity Science Center, Institute of Particle Physics, Hua-Zhong (Central China) Normal University, Wuhan 430079, China

E-mail: jiangj2007010209@gmail.com

Abstract. Inspired by the maxim "long union divides and long division unites", a phenomenological model with the simplification of real social networks is proposed to explore the evolutionary features of these networks composed of the entities whose behaviors are dominated by two events: union and division. The nodes are endowed with some attributes such as identity, ingredient, richness, age and internal diversity, which determine collectively the evolution in a probabilistic way. Through the local interaction of two events, a stationary state of network is reached as a constant amount of nodes survive with no more event happened in the network, like a situation of tripartite confrontation. Besides, the number of survived nodes and the speed of network evolution can be controlled by two parameters.

1. Introduction

This work is a network modeling of the social systems composed of a large number of entities in interaction whose existence is dominated by two major events: union and division. Union means the unification of two or more nodes (of the network) into one. Division is the inverse process: one node splits into several ones. Union and division are also the most visible social, economic, cultural and political phenomena in the course of the development of many composite systems. Different countries, political or economic groups can be unified into one country or group like the reunion of East and West Germany in 1990. There are also plenty of examples of division of these unities like the disintegration of Soviet Union in 1991. All of them are the consequence of the interplay of a pair of opposite tendencies in the evolutionary systems, and the recurrent character of these phenomena is well summarized in one Chinese maxim "long union divides and long division unites".

Union and division are veritable complex processes in which many factors are responsible for the consequences. A good example of this complexity is the evolution of different economic groups in the same field through the interaction including communication, influence, competition or cooperation etc. The evolution is composed of unification and splitting of groups and is under very complicated and uncertain influence of economic, political, scientific and technological systems as well as of many accidental elements such as natural disasters, wars, environmental changes and so on. It is for this reason that the most suitable description of the stochastic evolution of this kind of systems is the probabilistic modeling, taking into account as many of the involved factors and interaction mechanisms as possible.

Here, we present a phenomenological modeling of networks in which union and division of nodes are two main events determined in a probabilistic way by the nature and some general features of the nodes. One of the aims is to see what would be the destiny and the evolution characters of a network composed of nodes endowed with some general attributes allowing and influencing the union and division under some given conditions, without entering into the fundamental principles, interactions and true mechanisms of the dynamics of the systems. The definition of a network in the present work is inspired by the previous works on cultural networks [1, 2, 3, 4, 5, 6, 7, 8, 9, 10], where a culture is represented by a vector whose components are quantified by a limited number of values characterizing the cultural features. The vectorial representation of the nodes has been used in many other systems. For examples, in an economic network formed by companies (nodes), each company is labelled with several essential features like the richness or capital, the scale, the business scope, the survival time or brand image, and so on. In a political network composed of parties or groups, each group has its own political standpoint. In addition, the phenomenological model of union and division has been applied recently to a dual modeling of political opinion networks [11], in which union and division are two events happening in the subnetwork of political parties.

In the present model, we always have in mind some questions about, for instance: whether or not it is possible for all nodes of a given network to merge into one node under reasonable conditions. Or what will be the equilibrium or stationary state of network in which the number of nodes is relatively unchangeable, like the mode of tripartite confrontation? And what are the necessary conditions, if any, for these evolutionary behaviors? The reasons for considering these questions are that the globalization of economy, culture and politics is becoming an increasingly hot topic and raises various opinions in the world [12, 13, 14]. In our previous work [15], under some given conditions and definitions, we got some unexpected answers to these questions that the evolution of the system either terminated in a final union with the disappearance of the whole network and only a single node survived, or never ceased a nonequilibrium and periodic evolution with fluctuating oscillation of network size and some other quantities. Here, in order to keep finding the solution of the problems unsolved in our previous work, we extend our work with more realistic considerations. The main modifications include the redefinition of the quantity *richness* and the reintroduction of new rules of the merging and splitting processes, which results in some new interesting findings.

Again, we would like to emphasize that in this probabilistic model, the mechanism of union and division is not imposed deterministically on the evolution. At each step of evolution, union or division may or may not happen, depending on the probability of occurrence (union, division or nothing) calculated according to the nature and features of the nodes.

2. The model

Each node in the model represents a social entity such as a country, a culture, a political party, or a company. A link between two nodes means the interaction between them. So in one given network, each node denotes an individual entity when the object of entity is chosen, like in the company network, different nodes denote different companies with same business scope; in the culture network, nodes denote various cultures in the world. As stated above, some inherent properties could be used to characterize the entity. One can label a country or a culture by using some features like the language, religion, art, customs, economy, etc. A quantity named "identity" is introduced for this purpose and each node is endowed with an identity that has a position in the w dimensional space: $\vec{c} = (c_1, c_2, c_3, \dots c_w)(p \leq c_i \leq q)$ with $i = 1, 2, \dots w$ where w is the number of features and we consider $w = 3$ in the simulation for simplification. c_i is a random integer uniformly distributed, here we take $p = 1, q = 10$. The identity distance between two nodes i and j is defined as:

$$d_{ij} = \sqrt{\sum_{k=1}^{3} \Delta c_k{}^2}, \tag{1}$$

where $\Delta c_k = c_{ik} - c_{jk}$. This distance represents the difference between entities and similar entities have small distance between them.

As different entities in the system considered have in general different levels and power of development, so the quantity "richness" is introduced to be an important attribute of an entity. For a culture, this may stand for the size and the power of its content (religion, art, language, literature, economy, education, science and technology, etc.), the population of its carriers and so on. For a country, it could indicate economic, political, cultural, military power, living conditions, as well as the population, etc. A larger value of richness means a higher level of development and power. Compared to the time evolution of richness defined as the product of two attributes, the ingredient and age in Ref.[15], here, we reconsider and generalize it as follows:

$$\vec{r_i}(t) = \alpha_1 r_i(t)\vec{1} + \alpha_2 r_i(t)\vec{2} + \cdots + \alpha_w r_i(t)\vec{w}, \tag{2}$$

where α_i $(i = 1, 2 \ldots, w)$ is the weight of i-th feature and $\sum_{i=1}^{w} \alpha_i = 1$. This form is inspired by the concept of the rankings in society. For example, for an university, the University Rankings Index is one potential way to directly reflect the development or richness of one university. Like the famous Times University Ranking, five factors ($w = 5$) are contained as teaching, research, citations, industry income, and the international outlook with weight of 0.3, 0.3, 0.3, 0.025 and 0.075, respectively, and Harvard University had the highest richness value among 200 universities with top index 96.1 in the ranking 2010 [16]; for a company, the financial power (or turnover, wealth) could be labelled as richness for simplification with $w = 1$ such as Fortune 500 Index and the company "Wal-Mart Stores" got the highest richness value with the turnover 421849 million dollars among the Global Top 500 Companies in 2010 [17]; for a political party, the public support rate might be a suitable candidate for its richness. So various rankings of different entities like university, company, city competitiveness, etc, to some extent, could play the role of function of the richness in this model. Here, we take into account the age (a), the degree (k), and the ingredient (I) in the definition of richness for simplification as follows:

$$\vec{r_i}(t) = \alpha_1 a_i(t)\vec{1} + \alpha_2 k_i(t)\vec{2} + \alpha_3 I_i(t)\vec{3}, \tag{3}$$

where α_i $(i = 1, 2, 3)$ is weight value, and the age of a node is the time period from its birth to the present time, and the degree k or number of connections describes the communication state, the openness, the ability and will be of giving or receiving information of the node, and the ingredient I of a node is defined as the number of nodes that have been merged.

Next, we introduce two main dynamic processes in the network: merging and splitting. For two nodes randomly chosen, i and j , j is one of the nearest neighbors of i, they can merge into one new node with probability p_m proportional to the similarity or overlap (number of common identity features c_{li}, $l = 1, \cdots w$) between the identity of two nodes. The probability is defined as:

$$p_m(i, j) = \frac{1}{w} \sum_{l=1}^{w} \delta(c_{li}, c_{lj}), \tag{4}$$

where the δ function takes value 0 or 1 when c_{li} is equal or not equal to c_{lj}. This expression pays more attention to the similarity between the identity of two nodes since on the intuitive level, the more similar two nodes are, the more likely they are close to each other and merge together, as being said: *Birds of a feather flock together.* For example, two kinds of western or eastern cultures have more likely to merge than one eastern culture and one western culture do. If two nodes have totally different or identical features of identity , the probability of their merging will be zero or one. Otherwise, it will be $1/w \leq p_m < 1$. After the merging of two nodes, the new node n will be created with age $a = 1$ as a composite one characterized by the ingredient I. Obviously, the maximum ingredient is equal to the initial size of network N and will be conserved during the dynamic evolution. The richness of the new node evolves as Eq.(3) and its identity takes the average value of that of two merged nodes as following:

$$\vec{c} = \frac{c_{1i} + c_{1j}}{2}\vec{1} + \ldots + \frac{c_{wi} + c_{wj}}{2}\vec{w}. \tag{5}$$

The degree of the new node n is given by

$$k_n = k_i - 1 + k_j - 1 - N_{common}, \tag{6}$$

where N_{common} is the number of common neighbors of node i and j. In other words, the topological evolution of the network after node merging behaved as all neighbors of merged nodes i and j will be the neighbors of the new composite node n.

For the process of splitting in the model, a randomly selected composite node with identity I $(I \geq 2)$ could split in two ways. One is full or complete splitting generating I new nodes, has been studied in our previous work. The other proposed in the present work is partial splitting generating s new nodes $(2 \leq s \leq I - 1)$, in order to complete the mechanism of splitting and compare the effect on network dynamics caused by splitting methods. The splitting probability p_s is given from the new perspective of internal diversity as follows:

$$p_s = \frac{1}{Z_s}\overline{d_I} = \frac{1}{Z_s}\frac{2\sum_{i,j\in I} d_{ij}}{I(I-1)}, \tag{7}$$

where Z_s is the normalization constant, normalized over all composite nodes in the network, $\overline{d_I}$ is the average identity distance among I nodes merged by the composite node, which is also named as internal diversity, and i, j are the indexes of I nodes. The larger internal diversity $\overline{d_I}$ of one composite node is, the more likely it will fragment. In other words, the more internal solidarity one group has, the more difficult it disintegrates. After splitting, the identity of the new created nodes has the same dimensional number and the same range as the split "mother" node, with randomly given values uniformly distributed. The age of new nodes is equal to one and the ingredient of them is $I = 1$ for full splitting or a random integer between 1 and $I - s$ for partial splitting with the conservation that the sum of s new nodes' ingredients should be equal to I. In addition, the richness of each new node also agrees with Eq.(3). The structure of the network after node splitting is that these new nodes are randomly connected with each other and with the neighbors of the split node with a constant probability $p_{rc} = 0.9$.

It is worth noticing two conditions of time evolution in our simulation: 1) at a given time step, a composite node which has just been formed or a new node which has just split from a composite node must not be considered as the candidate for another merging or splitting; 2) during the random selection of the nodes, if the chosen node has $I = 1$, the one and only possible event is merging. But if it is a composite one, merging and splitting are both possible. The decision will take place with the changeable "event probability" $mors$ for each possible event in the present model, which is more realistic than the fixed probability 0.5 considered in the previous work.

The initial condition of the simulation is a number $N = 150$ of nodes randomly connected with a constant probability $p_c = 0.25$. The nodes remaining unconnected will be selected randomly at each time step and connected to other nodes according to the principle of preferential attachment.

As the weight of features defined in richness in Eq.(3) is unknown, the same weight value $\alpha_i = 1/3$ ($i = 1, 2, 3$) is implemented in simulation. When $t = 1$, the age of each node is $a = 1$, and the ingredient $I = 1$.

3. Numerical results

Totally different from a quasi-periodic behavior and an alternative replacement of union and division state found in the previous result, an equilibrium state is shown in the present model where the final size of network is stationary with time evolution. After the competition of union and division, the final size of the network is similar with the mode of tripartite confrontation and no more dynamic evolution happens. Details of the time evolution of some quantities are described below.

3.1. Cumulative network size evolution

The time evolution of cumulative network size with two splitting methods and different event probability $mors$ is shown in Fig. 1. The cumulative network size at time step t is defined as $\frac{\sum_{i=1}^{t} n(i)}{t}$, where $n(i)$ is the size of network (the number of nodes in network) at time step i. Since the final time step is various for network size evolution at stationary state with different event probability $mors$, the X-axis time scale of the figure is rescaled with the same range from 0 to 1 by dividing the final time step. In the left side of Fig.1, with partial splitting process, the cumulative network size decreases with the time growing for each event probability $mors$ with small variation. The merging and splitting processes are responsible for the oscillatory behavior of the cumulative size evolution. With the increase of event probability $mors$ from 0.1 to 0.9, or the decision to split for a composite node assumed in the simulation, the final network size increases from 15 to 36 approximately and the average final network size over different event probability $mors$ is around 26. In addition, the time step needed to reach the stationary state is sharply shortened with the increase of $mors$ during the evolution. In the right side of Fig.1, with full splitting process, the cumulative network size has similar tendency in time, but with large variation. The fluctuating range of final network size becomes much narrower between 24 and 34 approximately with event probability varying from 0.1 to 0.9 and the average final network size over different event probability is around 27.3 that is larger than that with partial splitting. Furthermore, the whole network needs more evolution time to reach the stationary state as more new nodes are split from the composite node with full splitting method compared to the case with partial splitting.

From the comparison between two splitting methods, we can draw a conclusion that the network can evolve to stationary state more quickly with partial splitting method, where larger stable network size could be found if full splitting method is used.

3.2. Richness distribution

We studied the richness distribution (represented by empty circle) and made a comparison with empirical data in Fig.2 in which the richness is normalized by dividing the maximum value of each data. The data sets include (1) the Global Urban Competitiveness Rankings Index 2007/2008 [18](data 1, represented by red circle): for each city competitiveness analysis over 500 main cities in the world, it is presented for seven sectors of the economy, such as industrial structure, human resources, the living environment, and so on. Each of these sectorial indices is the result of data for four to seven variables relating to aspects of each sector; (2) FT (full time) Weekly Median Pay (dollar per week) in 378 cities of USA in 2009 [19](data 2, represented by black square); (3) the World City Knowledge Competitiveness Index in 2008 [20](data 3, represented by blue square): it compares 145 regions across 19 knowledge economy benchmarks and five factors are selected for the analysis including human capital, knowledge capital, regional

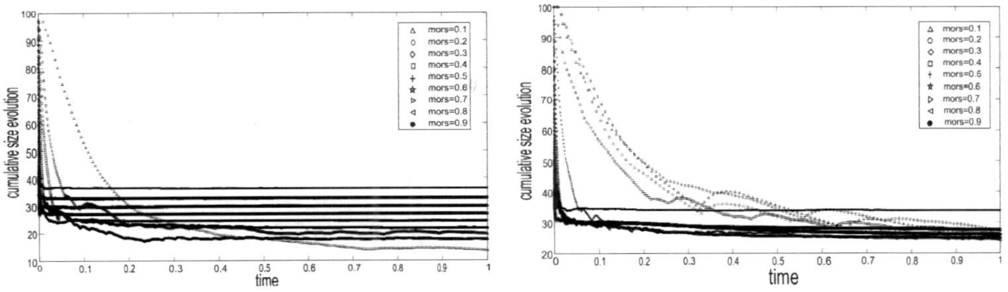

Figure 1: The evolution of cumulative size of network with different splitting processes and event probability $mors$. In the left side of figure, the cumulative size of network with partial splitting method decreases with the time evolution up to one relatively stationary value with tittle variation. With the increasing of $mors$ from 0.1 to 0.9, the final stable size of network increases from 15 to 36 approximately and less time step is needed to reach the stationary state. In the right side of figure with full splitting, similar behavior of cumulative size evolution is found over the time. The main differences from the left side are: 1) the fluctuating range of final stable size of network becomes narrow between 24 and 34 with event probability $mors$ varying from 0.1 to 0.9; 2) more time steps are needed for network to reach the stationary state with the increase of event probability.

economy outputs, financial capital and knowledge sustainability. As seen from the Fig.2, each distribution has a sharp peak at the point smaller than each richness mean value and the number of nodes with large richness is very small. The similarity between the simulation and empirical results suggests that the mechanism behind the definition of richness is reasonable and may have some connection with the nature of real complex system like the ranking system. In addition, the nature of richness also hints the fact that most of entities have low level of development or power, only a few of them can occupy the top level.

3.3. Distribution of the split nodes number

From the introduction of node splitting process in Section 2, we know that for the full splitting process, the number of split nodes from the composite node is equal to the ingredient I of composite node, and for the partial splitting process, the number of split nodes is between 2 and $I-1$. We record the number of split nodes during the whole evolution with given event probability $mors = 0.5$ in the simulation and show its probability distribution with two node splitting methods in Fig. 3. We find that 1) two distributions are power laws with fat tails in log-log plot, which suggests that during the time evolution, most of composite nodes with small ingredients split, and there are only a few of them with large ingredient, since these nodes have large probability to split. In the meanwhile, it also indicates that the probability distribution of ingredient matches the power law distribution; 2) power laws have two scaling exponents -1.4 and -1.6, fitted by two straight lines, corresponding to full and partial splitting method, respectively. Since for one composite node, partial splitting always produces less new nodes than full splitting does, so the scaling exponent for partial splitting is smaller than that for full splitting, and the maximum number of split node for former is also smaller than that for latter.

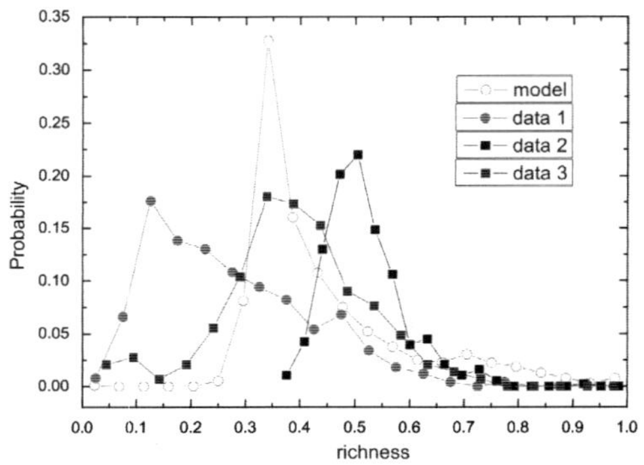

Figure 2: Comparison of the richness distribution between the model simulation and the real data including Top 500 Global Urban Competitiveness Rankings 2007/2008 (red circle, data 1)[18], the FT (full time) Weekly Median Pay in 378 cities of USA in 2009 (black square, data 2)[19], and the World Knowledge Competitiveness Rankings in 2008 of 145 cities in the world (blue square, data 3)[20]. We find that the model and real data have the similar tendency that each distribution has a sharp peak at the point smaller than each richness mean value, and most of entities are in the low level of development: only a few of them develops well and have much power.

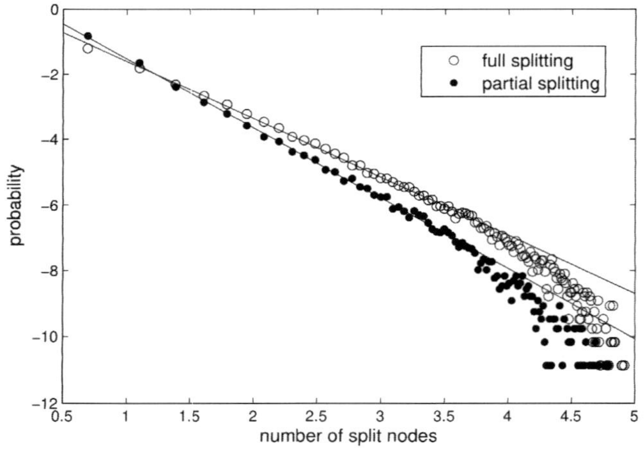

Figure 3: Distributions of the number of nodes split from composite node with full (empty circle) and partial (full circle) splitting methods. In log-log plot, distributions have the form of power law with fat-tail, with exponents -1.4 (full splitting) and -1.6 (partial splitting) fitted by two straight lines, respectively.

4. Conclusions and discussions

In the present paper, we have extended the phenomenological model of network evolution in our previous work with some important modifications including the redefinition of richness and the evolution rule of merging and splitting. We generalized the definition of richness and found an analogy between it and the ranking index in real system through the comparison between the simulation of richness distribution and some real data. In addition, inspired by the saying that birds of a feather flock together and the view of the internal diversity, we proposed new mechanisms of merging and splitting evolution functions.

As a result, some results totally different from previous work were found: for example, a stationary or equilibrium state of network size evolution has been obtained, in which a small amount of nodes survive in the network and no more merging or splitting process happens. The number of survived nodes depends on the value of event probability. Besides, the speed of network evolution is dependent on the node splitting method, full or partial splitting. At last, we found that the distribution of the number of node split from composite nodes follows the power law form with two splitting methods.

We want to stress that this model is still in its infancy and its applicability is limited with the social network where only the local interaction is dominated in the evolution, and more empirical data or accurate theoretical analysis is needed to support the simulation results. Thus this model has large flexibility to improve in the future work. For instance, global interaction or noise like random influence may take place in the process of merging and splitting and the features of some other quantities like the ingredient, the age and degree could be explored to enrich the findings of the model. We hope that this model, in spite of the simplification of the real world, could be a starting point for further work with more realistic and applicable considerations.

Acknowledgments

This work was supported by the National Science Foundation of China under grant No. 11405118.

References

[1] Axelrod R 1997 The dissemination of culture: A model with local convergence and global polarization, *J. Conflict Res* **41** 203-26.
[2] Stanoev A, Smilkov D and Kocarev L 2011 Identifying communities by influence dynamics in social networks *Preprint* arXiv:1104.5247v1.
[3] Palla G, Dereny I, Farkas I and Vicsek T 2005 Uncovering the overlapping community structure of complex networks in nature and society *Nature* **435** 814.
[4] Rosvall M and Bergstrom C T 2008 Maps of random walks on complex networks reveal community structure *Proc. Natl Acad. Sci.* USA **105** 1118.
[5] Clauset A, Newman M E J and Moore C 2004 Finding community strcutre in very large networks *Phys. Rev.* E **70** 066111.
[6] Lancichinetti A and Fortunato S 2011 Limits of modularity maximization in community detection *Preprint* arXiv:1107.1155v1.
[7] Newman M E J and Girvan M 2004 Finding and evaluating community structure in networks *Phys. Rev.* E **69** 026113.
[8] Newman M E J 2006 Modularity and community structure in networks *Proc. Natl. Acad. Sci.* USA **103** 8577.
[9] Kim B J, Trusina A, Minnlagen P and Sneppen K 2005 Self organized scale-free networks from merging and regeneration *Eur. Phys. J.* B **43** 369.
[10] Alava M J and Dorogovtsev S N 2005 Complex networks created by aggregation *Phys. Rev.* E **71** 036107.
[11] Wang R and Wang Q A 2011 Dual modeling of political opinion networks *Phys. Rev.* E **84** 036108.
[12] Cornwell G H and Stoddard E W 2001 Global Multiculturalism: Comparative Perspectives on Ethnicity, Race and Nation (Rowman & Littlefield Publishers).
[13] Falk R 1999 *Predatory Globalization: A Critique* (Prnceton: Prnceton University Press)
[14] Chang Y F and Cai X 2007 Macro-Behaviour of agents' opinion under influence of an external field *Chin. Phys. Lett.* **24** 2430.
[15] Jiang J, Wang R, Michel P and Wang Q A 2013 Long division unites or long union divides: A model for social network evolution *Chin. Phys. Lett.* **30** 3 038901.
[16] http://www.timeshighereducation.co.uk/world-university-rankings/
[17] http://www.fortunechina.com/fortune500/c/2011-07/07/content_62335.htm
[18] Ni P F and Kresl P K 2010 global urban competitiveness *report* 69-80.
[19] http://www.cforic.org/downloads.php
[20] Huggins R, Izushi H, Davies W and Shougui L 2008 World knowledge competitiveness index 7-13.

Least action principle and stochastic motion : a generic derivation of path probability

Aziz El Kaabouchi[1] **and Qiuping A. Wang**[1,2]
[1]ISMANS, LUNAM Université, 44, Avenue, F.A. Bartholdi, 72000, Le Mans, France.
[2]IMMM, UMR CNRS 6283, Université du Maine, 72085 Le Mans, France

E-mail: awang@ismans.fr

Abstract. This work is an analytical calculation of the path probability for random dynamics of mechanical system described by Langevin equation with Gaussian noise. The result shows an exponential dependence of the probability on the action. In the case of non dissipative limit, the action is the usual one in mechanics in accordance with the previous result of numerical simulation of random motion. In the case of dissipative motion, the action in the exponent of the exponential probability is just the one proposed in a previous work (Q.A. Wang, R. Wang, arXiv:1201.6309), an action defined for the total system including the moving system and its environment receiving the dissipated energy. In both cases, the result implies that the most probable paths are the paths of least action which, in the limit of vanishing randomness, become the regular paths minimizing the action.

1. Introduction

A characteristic of random motion is the existence of many possible paths between two given positions. A trivial experiment to show this is to drop a small feather in the air from a top point to a fixed point below. The first time it goes down along a zigzag path, the second time another path, and so on. If each path is defined as a band with a reasonable thickness instead of a single geometrical line, it is possible to assign a probability of occurrence to each of these paths, or the probability for a particle to take. In principle, this probability can be experimentally measured by observing a large number of identical feathers moving down between the two points, and estimated by the number of feathers observed along this path, divided by the total number of feathers arriving at the end point below. In practice, this kind of experiments is not that easy because, in order to get reliable calculation of probability, one has to count a very large number of identical bodies moving between two points under identical conditions.

In a recent work[1], we have proposed numerical measure of this probability by simulating the random motions subject to conservative forces. The aim of this probability measure is to determine its random variables. The random motion was generated by a Langevin equation with Gaussian noise. This numerical model is different from the above example of feathers in that there is no friction or energy dissipation on the falling bodies. This is an ideal motion or an asymptotic model for weakly damped random motion. The choice of this model, fully discussed in [1], is motivated by a theoretical study of path probability in the framework of a probabilistic Hamiltonian mechanics based on a generalization of least action principle of the regular Hamiltonian mechanics[3, 4, 5]. According to this theory, if the system is perturbed by a Gaussian white noise and statistically remains Hamiltonian one without or with very weak dissipative force, the path probability should be exponentially decreasing function $e^{-\gamma A}$ of the action $A = \int (K - V)dt$ where K is the kinetic energy, V the potential energy, γ a characteristic parameter of the randomness and the time integral is carried out along the considered path. This probability is a classical analog of the Feynman's factor $e^{\frac{i}{\hbar}A}$ of quantum Hamiltonian mechanics[2], the difference being that γ is a real number since the path probability must be real. The output of the numerical experiment has verified this theoretical prediction[1] and confirmed the generalization of the regular least action principle under these conditions.

A question following this work is whether or not the same exponential path probability exists for dissipative random motion. The answer to this question is crucial for the description of random motions in nature since most of them, if not all, are dissipative or over-damped. In this work, a analytical derivation of the path probability is provided first for the same dynamic model as in [1] in the zero dissipation limit and then for the case with important dissipation. The reader will find the above mentioned exponential path probability exists for both models. In the zero dissipation limit, the action is just the usual one as expected, while in the presence of important dissipation, the action is a new one suggested in a generalization of least action principle to dissipative motion[7]. This new action recovers the usual action when the dissipation tends to zero.

2. The model

The model can be depicted as the following Langevin equation

$$m\frac{d^2x}{dt^2} = -\frac{dV(x)}{dx} + f_d + R \tag{1}$$

where x is the one dimensional position, t the time, $V(x)$ the potential energy, f_d the friction force, and R the random forces as a white noise. A discrete solution of this equation is

$$x_i = x_{i-1} + r_i + f(t_i) - f(t_{i-1}) \tag{2}$$

which is a superposition of a Gaussian distributed random displacement r_i (solution of the stochastic equation $m\frac{d^2x}{dt^2} = R$) and a regular trajectory $f(t_i)$ (the solution of the Newtonian equation $m\frac{d^2x}{dt^2} = -\frac{dV(x)}{dx} + f_d$) corresponding to the least action path. Although Eq.(1) is not necessarily a linear equation in x, Eq.(2) is always its solution since the acceleration derivative $\frac{d^2}{dt^2}$ is a linear operator. Figure 1 illustrates some sample paths produced with Eq.(2) between two positions a and b, each of them being a tube of finite thickness δ (a band in the $x - t$ representation).

In the numerical experiment of the work [1], the number N_k of the particles (or geometrical lines) going through a given sample path k, i.e., all the sequences of positions $\{x_0, x_1, x_2 \cdots x_{n-1}, x_n\}$ satisfying $\{z_{k,i} - \delta/2 \le x_i \le z_{k,i} + \delta/2\}$ for all $i = 1, 2, \ldots, n$, is counted, where $\{a = z_{k,0}, z_{k,1}, z_{k,2}, \ldots, z_{k,n-1}, z_{k,n} = b\}$ are the axial line of that tube and n the total number of steps. The probability that the path k is taken is given by $P_k = N_k/N$ where N is the total number of particles (trajectories) moving from a to b through all the considered sample paths. The discrete normalization of P_k is $\sum_{k=1}^{s} P_k = 1$ where s is the total number of paths.

For each sample path, the instantaneous velocity of the particles of mass m at the step i is calculated by $v_{k,i} = \frac{z_{k,i} - z_{k,i-1}}{t_i - t_{i-1}}$ along the axial line. This velocity can be approximately considered as the average velocity of all the trajectories passing through the tube if δ is small with respect to the whole scale of the motion. The kinetic energy is given by $K_{k,i} = \frac{1}{2}mv_{k,i}^2$. The action along a path k is given by $A_k = \sum_{i=1}^{n} [\frac{1}{2}mv_{k,i}^2 - V(z_{k,i})] \cdot \Delta t$.

The Gaussian distribution of r_i at the step i is given by

$$p(r_i, \sigma) = \frac{1}{\sqrt{2\pi}\sigma} e^{-\frac{r_i^2}{2\sigma^2}}, \tag{3}$$

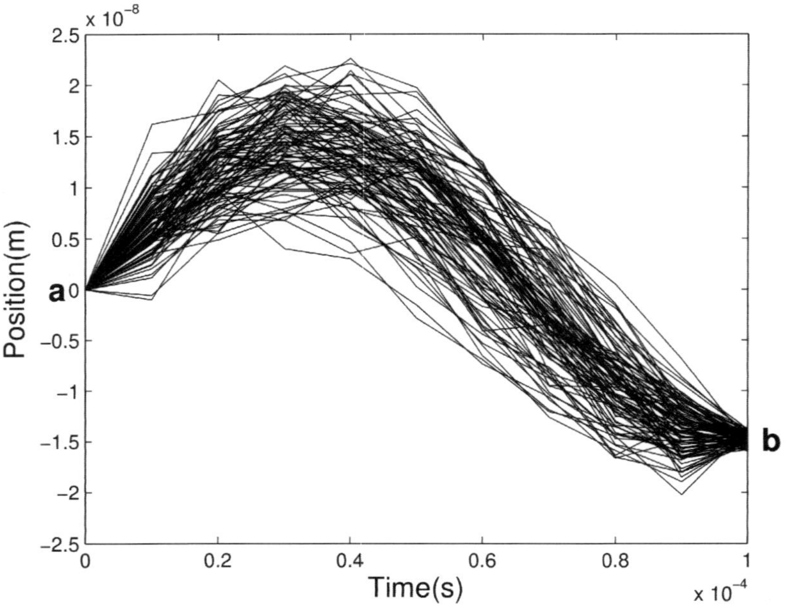

Figure 1. Sample paths created with Eq.(2) for a random motion subject to the Harmonic force and the Gaussian noise.

where, in the case of a free Brownian particle for instance, $\sigma = \sqrt{2D(t_i - t_{i-1})} = \sqrt{2D\Delta t}$ is the standard deviation, D the diffusion constant, and $\Delta t = t_i - t_{i-1}$ is the time interval of each step of the discrete motion.

3. Path probability of random motion in the non dissipative limit

This is the case where, in the Eq.(1), the dissipative force f_d is very weak with respect to the conservative force $-\frac{dV(x)}{dx}$, or the dissipative energy by the force f_d is small compared to the variation of the total mechanical energy.

From the definition of the Wiener measure[6], the probability P_k of a given path k can be expressed as follows:

$$P_k = \frac{\prod_{i=1}^{n} p_{k,i}}{\sum_{k=1}^{s} \prod_{i=1}^{n} p_{k,i}} \tag{4}$$

where $p_{k,i} = p(z_{k,i} - \delta/2 \leq X_i \leq z_{k,i} + \delta/2)$ for all $i = 1, 2, \ldots, n$, is the Gaussian distribution of all the realization x_i of the random variable X_i that are found in the tube of the path k. From Eq.(2), $p_{k,i}$ is given by

$$p_{k,i} = \frac{1}{\sqrt{2\pi}\sigma} \int_{z_{k,i}-\delta/2}^{z_{k,i}+\delta/2} \exp\{-\frac{[u - x_{i-1} - f(t_i) + f(t_{i-1})]^2}{2\sigma^2}\} du \tag{5}$$

For small δ, we have

$$p_{k,i} \approx \frac{\delta}{\sqrt{2\pi}\sigma} \exp\{-\frac{[z_{k,i} - x_{i-1} - f(t_i) + f(t_{i-1})]^2}{2\sigma^2}\} \tag{6}$$

This leads to

$$P_k \approx \frac{1}{Z}\exp\{-\frac{1}{2\sigma^2}\sum_{i=1}^{n}[z_{k,i} - x_{i-1} - f(t_i) + f(t_{i-1})]^2\} \tag{7}$$

where the normalization constant is given by

$$Z = \sum_{k=1}^{s}\exp\{-\frac{1}{2\sigma^2}\sum_{i=1}^{n}[z_{k,i} - x_{i-1} - f(t_i) + f(t_{i-1})]^2\}. \tag{8}$$

This implies

$$\ln P_k \approx -\frac{1}{2\sigma^2}\sum_{i=1}^{n}[z_{k,i} - x_{i-1} - f(t_i) + f(t_{i-1})]^2 - \ln Z. \tag{9}$$

As Z is constant, the function $\ln P_k$ depends only on the sum over i. Let it be S_k, we have

$$\begin{aligned} S_k = &\sum_{i=1}^{n}(f(t_i) - f(t_{i-1}))^2 + \sum_{i=1}^{n}(z_{k,i} - z_{k,i-1})^2 + \sum_{i=1}^{n}(z_{k,i-1} - x_{i-1})^2 \\ &+ 2\sum_{i=1}^{n}(z_{k,i} - z_{k,i-1})(z_{k,i-1} - x_{i-1}) \\ &+ 2\sum_{i=1}^{n}(z_{k,i-1} - x_{i-1})(f(t_i) - f(t_{i-1})) \\ &- 2\sum_{i=1}^{n}(z_{k,i} - z_{k,i-1})(f(t_i) - f(t_{i-1})) \end{aligned} \tag{10}$$

We remark that :

- The first term is constant because it is a sum over the deterministic path $f(t)$.
- The second term in Eq.(5) is just the time sum (time integral in continuous version) of the kinetic energy, i.e., $\sum_{i=1}^{n}(z_{k,i} - z_{k,i-1})^2 = \frac{2\Delta t}{m}\sum_{i=1}^{n}\frac{1}{2}mv_{k,i}^2\Delta t$.
- The third term $\sum_{i=1}^{n}(z_{k,i-1} - x_{i-1})^2 \leq n\delta^2$ can be neglected for small δ.
- The fourth term is such that

$$2\sum_{i=1}^{n}(z_{k,i} - z_{k,i-1})(z_{k,i-1} - x_{i-1}) \leq 2\delta n \max_{\substack{k=1,\ldots,s \\ i=1,\ldots,n}} |z_{k,i} - z_{k,i-1}|\,. \tag{11}$$

It follows that this term can be neglected for small δ because $\max_{\substack{k=1,\ldots,s \\ i=1,\ldots,n}} |z_{k,i} - z_{k,i-1}|$ is constant.

- The fifth term is such that

$$2\sum_{i=1}^{n}(z_{k,i-1} - x_{i-1})(f(t_i) - f(t_{i-1})) \leq 2\delta n \max_{i=1,\ldots,n} |f(t_i) - f(t_{i-1})|\,. \tag{12}$$

Consequently, this term can be neglected for small δ because $\max_{i=1,\ldots,n} |f(t_i) - f(t_{i-1})|$ is constant.

• Finally, the sixth term in Eq.(5) reads

$$\begin{aligned}-2\sum_{i=1}^{n}(z_{k,i}-z_{k,i-1})(f(t_i)-f(t_{i-1})) =& 2\sum_{i=1}^{n-1} z_{k,i}\left(f(t_{i-1}-2f(t_i)+f(t_{i+1})\right)\\ &+2z_{k,0}[f(t_1)-f(t_0)]+2z_{k,n}[f(t_n)-f(t_{n-1})]\\ =&2\sum_{i=1}^{n-1} z_{k,i}(\Delta t)^2\left(\frac{\frac{f(t_{i+1})-f(t_i)}{\Delta t}-\frac{f(t_i)-f(t_{i-1})}{\Delta t}}{\Delta t}\right)\\ &+2a[f(t_1)-f(t_0)]+2b[f(t_n)-f(t_{n-1})]\end{aligned} \tag{13}$$

It follows that for (Δt) small, we have

$$\begin{aligned}-2\sum_{i=1}^{n}(z_{k,i}-z_{k,i-1})(f(t_i)-f(t_{i-1})) \approx& 2\sum_{i=1}^{n-1} z_{k,i}(\Delta t)^2\left(\ddot{f}(t_i)\right)\\ &+2a[f(t_1)-f(t_0)]+2b[f(t_n)-f(t_{n-1})]\end{aligned} \tag{14}$$

At this stage, S_k reads

$$S_k = \frac{2\Delta t}{m}\left[\sum_{i=1}^{n}\frac{1}{2}mv_{k,i}^2\Delta t+\Delta t\sum_{i=1}^{n-1} z_{k,i}m\ddot{f}(t_i)\right]+constant \tag{15}$$

where $constant$ includes all the constant terms independent of k mentioned above. Since $m\ddot{f}(t_i) = -\frac{dV(f(t_i))}{df(t_i)} = -V'(f(t_i))$, we get

$$S_k = \frac{2\Delta t}{m}\left[\sum_{i=1}^{n}\frac{1}{2}mv_{k,i}^2\Delta t-\sum_{i=1}^{n-1} z_{k,i}V'(f(t_i))\Delta t\right] \tag{16}$$

to a constant. In order to introduce the potential energy $V(z_{k,i})$ on the path k, we consider the following approximation

$$V(z_{k,i}) \approx V(f(t_i))+(z_{k,i}-f(t_i))V'(f(t_i)), \tag{17}$$

or $z_{k,i}V'(f(t_i)) \approx V(z_{k,i})-V(f(t_i))+f(t_i))V'(f(t_i))$, which leads to

$$S_k = \frac{2\Delta t}{m}\left[\sum_{i=1}^{n}\frac{1}{2}mv_{k,i}^2\Delta t-\sum_{i=1}^{n-1} V(z_{k,i})\Delta t\right] \tag{18}$$

to a constant including all the terms independent of of the path k in the Eq.(17). Let us introduce the usual action of the mechanics A_k given by

$$A_k = \sum_{i=1}^{n} L_{k,i}\Delta t \tag{19}$$

where the Lagrangian is $L_{k,i} = K_{k,i} - V(z_{k,i})$, the kinetic energy $K_{k,i} = \frac{1}{2}mv_{k,i}^2$, and the potential energy $V(z_{k,i})$, all along the path k, up to the time t_i. Finally, the probability of the path k is given by

$$P_k = \frac{1}{Z}\exp(-\eta A_k) \tag{20}$$

with $\eta = \frac{\Delta t}{m\sigma^2}$. All the terms independent of k vanish after the normalization over k with the constant $Z = \sum_{k=1}^{s}\exp(-\eta A_k)$. As η is always positive, Eq.(20) means that the most probable path must be the paths of least action A_k. This probability has been observed in the numerical experiment of random motion in the weak dissipative limit in the work [1].

4. Path probability of dissipative random motion

The usual action loses its role of characteristic variable for the paths in the variational calculus when dissipation is present. The search for an alternative variable has been since long a headache for physicists in analytical mechanics, as discussed briefly in [7].

In order to introduce the dissipative force f_d and the energy it dissipated in a Lagrangian to calculate the action, we consider the work E_d of f_d, along the path $f(t_i)$, from the time $t = t_0 = 0$ to the time $t = t_i$, i.e., $E_d(f(t_i)) = -\int_0^{f(t_i)} f_d(x)dx$. E_d, being the energy lost by the system to the environment, must be positive. We have, according to the second fundamental theorem of calculus, the following expression $f_d = -\frac{dE_d(f(t_i))}{df(t_i)} = -E_d'(f(t_i))$.

Now Eq.(16) can be rewritten with the help of the following equation of motion

$$m\ddot{f}(t_i) = -V'(f(t_i) - E_d'(f(t_i)) \tag{21}$$

and the approximation

$$\begin{aligned} V(z_{k,i}) + E_d(z_{k,i}) &\approx V(f(t_i)) + E_d(f(t_i)) \\ &+ (z_{k,i} - f(t_i))\left[V'(f(t_i)) + E_d'(f(t_i))\right] \end{aligned} \tag{22}$$

After arrangement, S_k reads, to a constant independent of the index k,

$$S_k = \frac{2\Delta t}{m} A_k \tag{23}$$

where the action A_k is given by

$$A_k = \sum_{i=1}^{n} L_{k,i}\Delta t \tag{24}$$

with the Lagrangian $L_{k,i} = K_{k,i} - V(z_{k,i}) - E_d(z_{k,i})$, the kinetic energy is $K_{k,i} = \frac{1}{2}mv_{k,i}^2$, the conservative potential energy $V(z_{k,i})$ and the energy dissipated $E_d(z_{k,i})$, along the path k up to the time t_i. Finally, the path probability of the path k is given by

$$P_k = \frac{1}{Z}\exp(-\eta A_k) \tag{25}$$

All the terms independent of k vanish after the normalization over k with the constant $Z = \sum_{k=1}^{s}\exp(-\eta A_k)$.

The quantity A_k is still called action for dissipative system because, in a variational calculus proposed in [7], its vanishing variation $\delta A_k = 0$ can lead to the Newtonian equation of motion with dissipative force. It is then a characteristic variable of paths for dissipative motion in the same way as the usual action for non dissipative motion of Hamiltonian systems. This 'dissipative' action recovers the usual action for the case where E_d is negligible with respect to the energy of the system.

5. Concluding remarks

In this work, the path probability for random dynamics of mechanical system has been calculated on the basis of a model described by Langevin equation with Gaussian noise. The result shows an exponential dependence of the probability on the action which confirms the measure of the path probability by numerical simulation of random motion with the same model in the non dissipative limit. In the case of dissipative motion, the action can be calculated with a Lagrangian which include the energy lost by the system due to the dissipation, which confirms a generalization of the principle of least action of Hamiltonian systems to dissipative systems[7]. According to the result of this work, for both non dissipative and dissipative systems, the exponential path probability implies that the most probable paths are the paths of least action which, in the limit of vanishing randomness, become the regular paths of least action.

References

[1] Lin T L, Wang R, Bi B W, El Kaabouchi A, Pujos C, Calvayrac F and Wang Q A 2013 Path probability distribution of stochastic motion of non dissipative systems: a classical analog of Feynman factor of path integral *Chaos, Solitons & Fractals* **57** 129
[2] Feynman R P and Hibbs A R 1965 *Quantum mechanics and path integrals* (New York: McGraw-Hill Publishing Company)
[3] Wang Q A 2004 Maximum path information and the principle of least action for chaotic system *Chaos, Solitons & Fractals* **23** 1253
[4] Wang Q A, Tsobnang F, Bangoup S, Dzangue F, Jeatsa A and Le Méhauté A 2009 Reformulation of a stochastic action principle for irregular dynamics *Chaos, Solitons & Fractals* **40** 2550
[5] Wang Q A 2005 Non quantum uncertainty relations of stochastic dynamics *Chaos, Solitons & Fractals* **26** 045
[6] Mazo R M 2002 *Brownian motion* (London: Oxford University Press)
[7] Wang Q A and Wang R Is it possible to formulate least action principle for dissipative systems? *Preprint* arXiv:1201.6309

Dynamics of rotator chain with dissipative boundary: energy conduction

Pu Ke and Zhigang Zheng
Department of Physics, Beijing Normal University, Beijing 100875, China
E-mail: zgzheng@bnu.edu.cn

Abstract. Dynamical states of coupled rotator lattice are identified and their effects on energy conduction are explored. Globally synchronous rotation, clustered synchronous rotation and split synchronous rotation states are observed and the heat baths induced transitions between these states causes the reduction of overall energy flux. This result provides an explanation to the finite heat conductivity of coupled rotator lattice model.

1. Introduction

Non-equilibrium transport process is a classical yet challenging problem of non-equilibrium statistical physics. The macroscopic description of non-equilibrium transport phenomena depends on transport equations and the core of all transport equations are the linear laws that depict the amount of extensive quantities which are transferred per unit time between different regions of a system as responses to non-homogeneities in intensive quantities.

The ultimate goal of non-equilibrium statistical physics is to derive these linear laws from rigorous microscopic arguments, which is of formidable difficulty. Since the most time tested transport law in experiments is the Fourier's law on energy transport $\mathbf{j}(\mathbf{x},t) = -\kappa(T)\nabla T(\mathbf{x},t)$, where $\mathbf{j}(\mathbf{x},t)$ is the energy flux, $T(\mathbf{x},t)$ is the local temperature and $\kappa(T)$ is the heat conductivity, many works have been done on the energy conduction properties. After the elementary attempt of Debye phonon gas theory, the Boltzmann-Peierls approach showed anharmonicity is necessary to give genuine diffusion of energy on a lattice through the so-called Umklapp processes[1]. However, there are more fundamental questions beyond the scope of the above theories such as the uniqueness of non-equilibrium stationary state and the sufficient condition for local equilibrium, which penetrate down to the level of dynamics. Thus great effort had been put into the investigation of transport properties on simple lattice models by simulating their dynamics in the hope of providing a more solid foundation to Fourier's law[2–17].

Contrary to the expectation of physicists back then, Fourier's law was found to be violated on many one dimensional lattice models by numerical simulations for heat conductivity diverges with a power law as the scale of the system increases[18, 19], i.e. $\kappa(N) \propto N^{\alpha}$.

These findings triggered the query of the necessary and sufficient condition for a system conforming the Fourier's law of heat conduction(i.e. whether heat conduction on a system is normal or anomalous). Decades of observations found decisive evidences that on-site potential, which breaks the momentum conservation of a lattice, is a sufficient condition for normal conduction, provided that the temperature is sufficiently low. Numerical simulations of Ding-a-ling model and Frenkel-Konorova model support the assertion[18, 20]. On top of that, analytical analysis even proved that for momentum conserving lattice Fourier's law does not hold, which implies that the non-conservation of momentum is a necessary condition for normal heat conduction[5]. Paradoxically, one system serves as an counter example of the above mathematically rigorous conclusion — coupled rotator lattice, without any on-site potential, exhibits normal heat conduction[21, 22]. Ergo, more detailed analysis need to be performed on the coupled rotator chain model to reconcile the contradiction between the analytical and numerical results.

The study of anomalous heat conduction in one-dimensional momentum-conserving systems was based on the oscillator lattices, whose main energy carriers are phonons and solitons. But in a coupled rotator lattice, non-linear rotating modes(i.e. breathers) also exist[23]. Thus the effects of rotating modes in the energy conduction on coupled rotator model were explored to probe their effects on heat conduction. Our explorations indicate that the non-linear rotating modes hamper the heat conduction, thus provide an explanation to the finite heat conductivity of coupled rotator chain model.

2. Coupled rotator lattices

Non-linear rotating modes could be excited by applying mechanical torque on a coupled rotator chain, which greatly facilitated our investigation, and the feasibility of this setting is grounded on the interaction potential of the model. The coupled rotator lattice model is the simplest classical 1D spin model called the Heisenberg XY model, whose interaction potential is sinusoidal coupling. Assuming the model is described by the angles $\boldsymbol{\phi} = (\phi_1, \phi_2, \ldots, \phi_N)$ and their conjugate angular momenta $\boldsymbol{L} = (L_1, L_2, \ldots, L_N)$, see FIG. 1., then the interaction potential

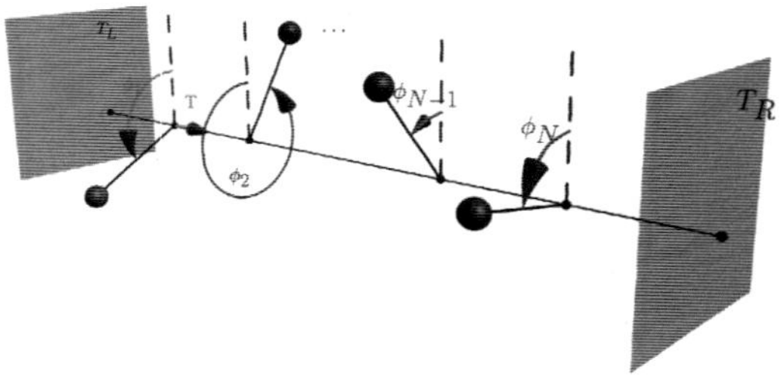

Figure 1. Illustration of a coupled rotator lattice with heat bath on each end, mechanical torque T can be applied on the system.

takes the form

$$U(\phi_k, \phi_{k+1}) = \epsilon\left[1 - \cos(\phi_{k+1} - \phi_k)\right], \tag{1}$$

where ϵ is the coupling coefficient. Note that unlike the oscillator based lattice models, the interaction potential is periodic, which enables both oscillations and relative rotations among the rotators on the lattice rather than only oscillations. On top of that, applied torque can also be added to the system to artificially excite rotational modes.

Moreover, the system is homogeneous, that is, the moment of inertia of each rotator is I. Therefore, the Hamiltonian of the system is

$$\mathscr{H}(\boldsymbol{\phi}, \boldsymbol{L}) = \sum_k \left\{ \frac{L_k^2}{2I} + \epsilon\big[1 - \cos(\phi_{k+1} - \phi_k)\big] \right\}. \tag{2}$$

The total angular momentum of this system is conserved, since $\dot{L} = \sum_k \dot{L}_k = \sum_k \partial\mathscr{H}/\partial\phi_k = 0$. If momentum conservation was the sufficient condition for the violation of Fourier's law, then the heat conduction on a coupled rotator lattice should be anomalous. However, independent researches concluded that the heat conduction of a coupled rotator lattice is normal when the temperature is sufficiently high[21, 22]. It is suggested that the excitation of stationary localized rotational modes on the coupled rotator lattice facilitate the locking of the heat flux hence render the heat conductivity finite.

The stationary localized rotational modes could be studied by adding mechanical torque to the system. Based on the previous researches one would expect Fourier's law is also valid in this setting. Nevertheless, according to the study conducted on a coupled rotator lattice with both thermal baths and mechanical torque exist on the boundary, Fourier's law was violated. Fourier's law implies the stationary local temperature distribution is linear, and the heat conductivity is positive. But the temperature distribution on a coupled rotator lattice was single peaked and the heat flux was reduced when the temperature difference was increased(i.e. the heat conductivity appears to be negative). It was asserted that the two counter-intuitive phenomena were the consequence of the interaction between thermal and mechanical driving(applied torque). More specifically, the reduction of heat flux was considered to be the result of the negative flux induced by the mechanical and thermal interaction[24]. Since applied torque could excite non-linear rotational modes, we intent to study their effects on the thermal transport to find the mechanism behind the single peaked "temperature" distribution and the reduction of "heat flux".

3. Propagation dynamics: energy transport

The incorporation of applied torque introduces mechanical energy flux, which is from different origin of heat flux, it is desirable to separate these two kinds of fluxes from a dynamical perspective. In order to achieve that objective, a reductionism approach was utilized: we first investigated the energy conduction properties of a coupled rotator lattice without heat baths, then the interactions of heat baths and applied torque were explored to study the effect of rotational mode on energy conduction.

3.1. Equation of motion, local temperature and energy flux

Having introduced the dimensionless time $s = \sqrt{\frac{\epsilon}{I}}t$ and letting $\omega_k = \frac{\mathrm{d}\phi_k}{\mathrm{d}s}$, the dimensionless equations of motion for a coupled rotator lattice with open boundary can be derived from (2)

$$\begin{aligned}
\frac{\mathrm{d}\phi_k}{\mathrm{d}s} &= \omega_k, \\
\frac{\mathrm{d}\omega_k}{\mathrm{d}s} &= \sin(\phi_{k+1} - \phi_k) - \sin(\phi_k - \phi_{k-1}), \qquad\qquad k \neq 1, N \\
\frac{\mathrm{d}\omega_1}{\mathrm{d}s} &= \sin(\phi_2 - \phi_1) - \alpha\omega_1 + \tau + \sqrt{\eta_1}\xi_1, \\
\frac{\mathrm{d}\omega_N}{\mathrm{d}s} &= -\sin(\phi_N - \phi_{N-1}) - \alpha\omega_N + \sqrt{\eta_2}\xi_2,
\end{aligned} \tag{3}$$

where $\alpha = \gamma\sqrt{\frac{I}{\epsilon}}$ is the effective dissipation coefficient, $\eta = \frac{2I^2\alpha T}{\epsilon^2}$ is the noise strength, and $\tau = \frac{\mathrm{T}}{\epsilon}$ is the effective torque, in which γ is the dissipation coefficient and T is the mechanical torque. Heat baths are modeled by Gaussian random force ξ.

Local temperature and local energy flux are common observables to investigate the relationship between the temperature gradient and the energy flux in the low-dimensional system.

The definition of temperature is borrowed from classical equilibrium statistical mechanics according to Equipartition theorem

$$\begin{aligned} E_{\text{total}} &= \frac{1}{2}\sum_i m_i(v_i' + v_c)^2 = \frac{1}{2}\sum_i m_i v_i'^2 + \frac{1}{2}Mv_c^2 \\ &= \frac{N}{2}k_B T + \frac{1}{2}Mv_c^2 \Rightarrow T = \frac{1}{k_B N}\sum_i m_i(v_i - v_c)^2, \\ &\Rightarrow T = \frac{1}{k_B N}\sum_i m_i(v_i - v_c)^2, \end{aligned}$$

where v_i is the velocity of each particle and v_c is the velocity of the center of mass. By setting $m_i = 1$ and $k_B = 1$

$$T = \text{var}\{v\}, \tag{4}$$

which is the variance of velocity. Despite the constant factor 2 its physical meaning is straightforward — the mean internal energy. The corresponding local temperature for each rotator in a coupled rotator chain is then followed naturally

$$\begin{aligned} T_k &= \text{var}\{\omega_k\} \\ &= \lim_{s\to\infty}\frac{1}{s}\int_0^s \omega_k^2(t)\mathrm{d}t - \left[\lim_{s\to\infty}\frac{1}{s}\int_0^s \omega_k(t)\mathrm{d}t\right]^2. \end{aligned} \tag{5}$$

To derive the definition of local energy flux, one notes that energy could not be created or

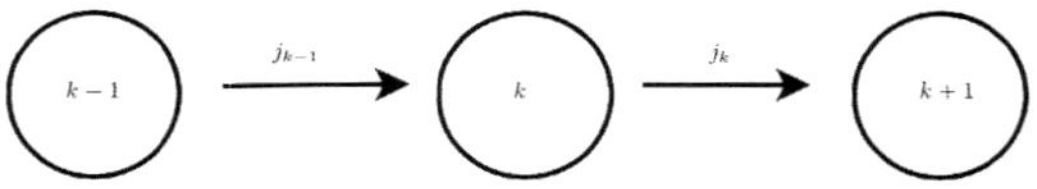

Figure 2. Illustration of energy flux

destroyed in the rotator chain, then the continuity equation holds

$$\frac{\mathrm{d}h_k}{\mathrm{d}s} = j_{k-1} - j_k, \tag{6}$$

where h_k and j_k are local energy and energy flux.

The expression of local energy flux could be obtained by comparing (6) with the dimensionless time derivative of effective local energy that is derived from the equation of motion. If one adopts the dimensionless coordinates introduced in (3), then the effective local energy is

$$h_k = \frac{\omega_k^2}{2} + \frac{1}{2}\Big[2 - \cos(\phi_{k+1} - \phi_k) - \cos(\phi_k - \phi_{k-1})\Big], \tag{7}$$

in which the first term on the right hand side is the effective kinetic energy and the second term is the effective interaction potential energy. Its dimensionless time derivative is

$$\frac{\mathrm{d}h_k}{\mathrm{d}s} = \omega_k\frac{\mathrm{d}\omega_k}{\mathrm{d}s} + \frac{1}{2}\Big[(\omega_{k+1} - \omega_k)\sin(\phi_{k+1} - \phi_k) + (\omega_k - \omega_{k-1})\sin(\phi_k - \phi_{k-1})\Big]. \tag{8}$$

Substituting (3) to (8)

$$\frac{\mathrm{d}h_k}{\mathrm{d}s} = \frac{1}{2}\Big[(\omega_k + \omega_{k-1})\sin(\phi_k - \phi_{k-1}) - (\omega_{k+1} + \omega_k)\sin(\phi_{k+1} - \phi_k)\Big], \tag{9}$$

and comparing the (9) to (6) leads to the definition of energy flux

$$j_k = \frac{1}{2}(\omega_{k+1} + \omega_k)\sin(\phi_{k+1} - \phi_k). \quad (10)$$

Dimensionless time average is used in the numerical calculation of energy flux instead of ensemble average,

$$\langle j_k \rangle = \lim_{s\to\infty} \frac{1}{2s}\int_0^s [\omega_k(t) + \omega_{k+1}(t)]\sin[\phi_{k+1}(t) - \phi_k(t)]\mathrm{d}t. \quad (11)$$

In steady states $\mathrm{d}h_k/\mathrm{d}s = 0$, hence the local energy flux on each rotator is equal. Not only is the average energy flux an important physical quantity for the properties of energy conduction, overall dynamics of rotator chain could also be reflected in the variation of energy flux.

One caution needs to be noted here, in numerical simulations we use (dimensionless) time average to calculate the quantities instead of ensemble averages. Previous study shows the ensemble averages and time averages agree with each other in weakly coupled rotator chains, thus the substitution should not raise any problem.

The system of equations (3) had been integrated numerically by Velocity-Verlet method for a chain of 1024 rotators with the time step size $\Delta s = 0.01$.

3.2. Dynamical states of coupled rotator lattice

To understand the mechanism behind the single peaked local temperature distribution and the reduction of energy flux, one needs to locate the dynamical states that cause the phenomena. Transitions between different dynamical states can be directly reflected in the qualitative change of averaged energy flux, thus averaged energy flux is calculated to reveal distinct dynamical states, see FIG. 3.

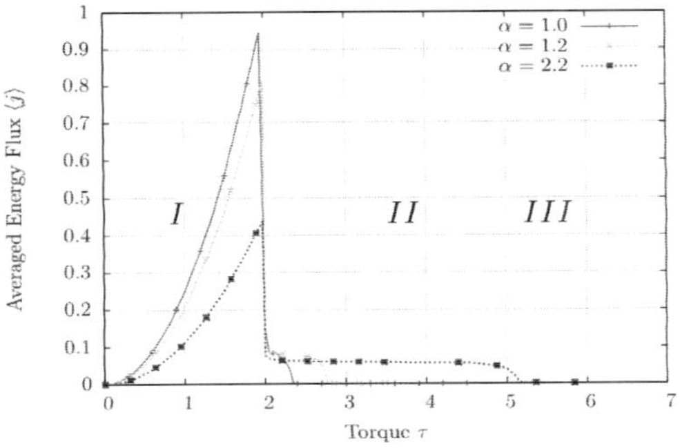

Figure 3. Averaged energy flux of a deterministic coupled rotator lattice, three distinct dynamical states were identified.

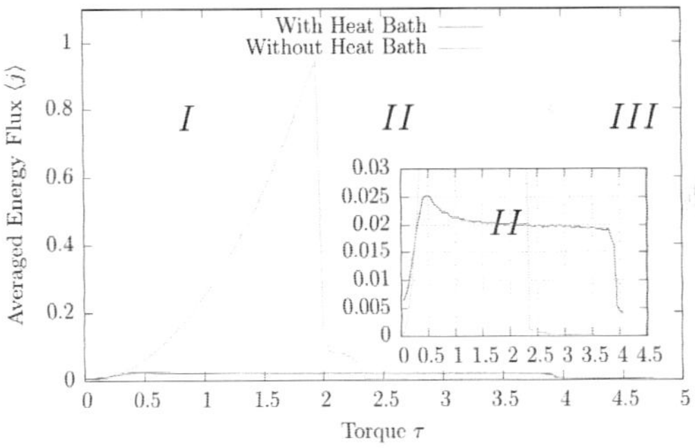

Figure 4. Averaged energy flux of a coupled rotator lattice with heat baths, the plateau was broadened.

The torque-flux curve is divided into three qualitatively different regions, which correspond to three distinct dynamical states. In region I, the local energy flux increases monotonically with driving torque; In region II, the curve plunges and exhibits a plateau; and in region III, the flux virtually vanishes. One would expect temperature gradient always enhances the energy flux, but that is not the case. The incorporation of thermal heat baths enhances the overall energy flux when applied torque is small, while reduces the flux when applied torque is large, the plateau is also broadened, see FIG. 4.

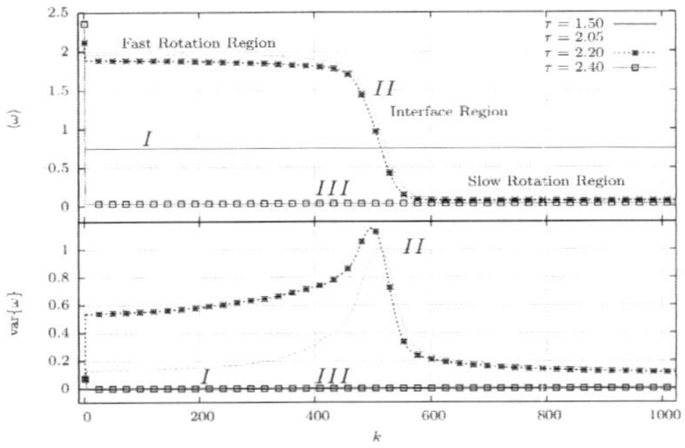

Figure 5. Averaged momenta and variance of momenta for chains of $\alpha = 1.0$.

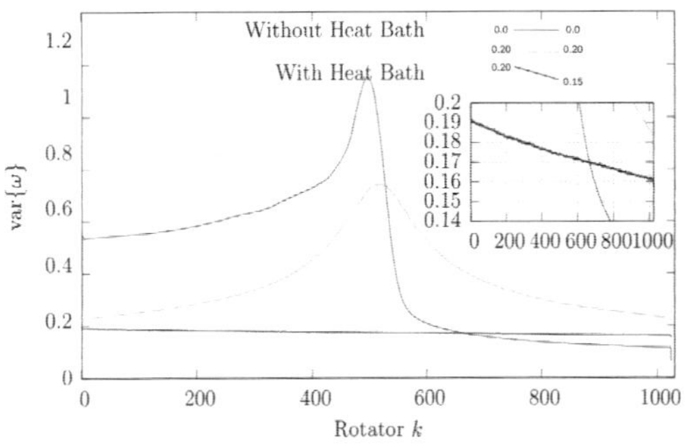

Figure 6. Comparison between a rotator chain with and without heat baths.

After identifying the three states, the dynamics of each state could be understood by the averaged angular momenta and momenta variance profile, see FIG. 5.

In region I, the entire lattice has the same averaged angular momentum and the variance is also identically zero, which indicates the lattice was synchronized, thus the dynamical state is called *globally synchronous rotation*. In region III, only the driven rotator has relatively large averaged angular momentum and non-zero variance, the other rotators are synchronized and rotates slowly, hence this state is named *split synchronous rotation*. The most intriguing dynamical state is II, in which the lattice is divided into three regions 1. fast rotation region 2. slow rotation region and 3. interface region. It is interesting to find that the single peaked distribution emerges even when the heat baths are absent. If heat baths can not destroy the single peaked distribution, then we can safely draw the conclusion that the distribution is of deterministic origin. The results of the simulation of a coupled rotator chain with both applied torque and heat baths confirms our hypothesis, see FIG. 6, hence the single peaked "temperature" distribution indeed results from deterministic dynamics of coupled rotator chain model.

Since the hump is located on the interface region, understanding the deterministic origin of the single peaked momenta variance distribution requires further investigations on the dynamics of interface rotators, thus the momentum distribution of interface rotators are calculated, see FIG. 7. Rotators in the interface region(450, 472 and 490) have momenta distributions extended

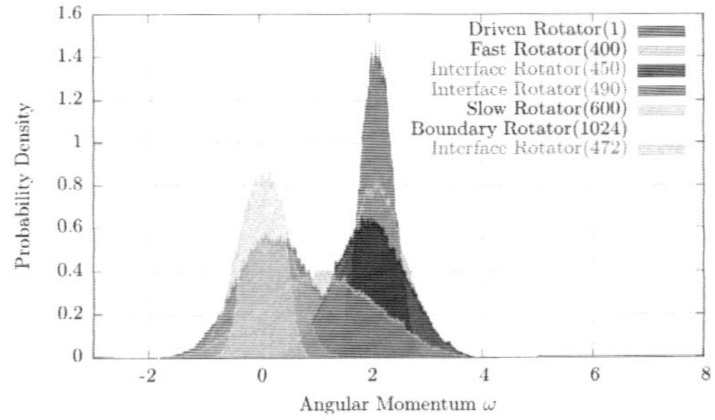

Figure 7. Distribution of angular momenta for $\tau = 2.2, \alpha = 1.0$.)

from the slow rotation region to the fast rotation region, which implies these interface rotators constantly switched between the slow rotation state and the fast rotation state, which result in large variance. This behavior leads to the name *clustered synchronous rotation*.

In summary, our research suggests that the mechanism of the peak in the temperature profile of a coupled rotator lattice is the torque-induced rotating interface. In the interface, rotators switch intermittently between the fast and slow rotations, which leads to the large variance of angular momenta. The local energy flux is also reduced after the system enters the region II and region III, and the incorporation of heat baths broaden the region II.

3.3. Dynamical states of coupled rotator

Since the non-linear rotation modes is induced by the applied torque on the boundary, it is desirable to reduce the system to the utmost degree to get a deeper understanding of the interaction of thermal driving and mechanical driving. Thus a coupled rotator that consists of only two rotators is investigated analytically to seperate the contributions of heat baths and the applied torque to the overall energy flux and to provide insight into the mechanism behind the reduction of energy flux.

It can be proven that a deterministic coupled rotator has two dynamical states, synchronized rotation and unsynchronized rotation[25], which are divided by $\tau = 2$. The unsynchronized rotation corresponds to the non-linear rotation mode where relative rotation exist. The local energy flux can be easily calculated by

$$\langle j \rangle = \begin{cases} \frac{\tau^2}{4\alpha}, & \text{synchronized rotation} \\ \frac{\tau(\tau-\sqrt{\tau^2-4})}{4\alpha}, & \text{unsynchronized rotation} \end{cases} \tag{12}$$

which perfectly matches the numerical simulation, see FIG. 8. The local energy flux plunges immediately after the system enters the unsynchronized rotation state, which is consistent with our prediction. When heat baths is incorporated into the system, directly solving the related

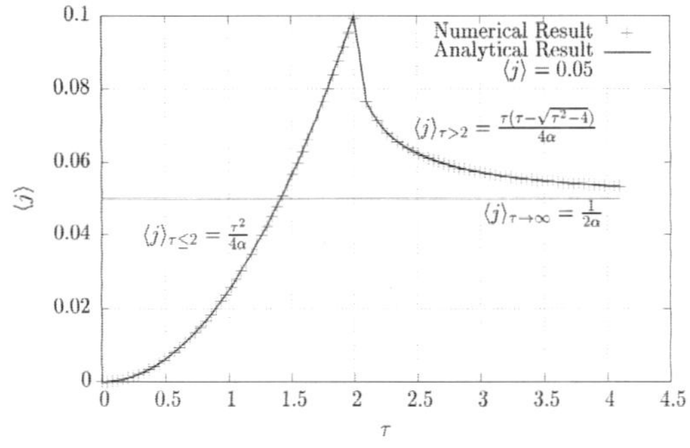

Figure 8. The energy flux curve of a coupled rotator, $\alpha = 10.0$.

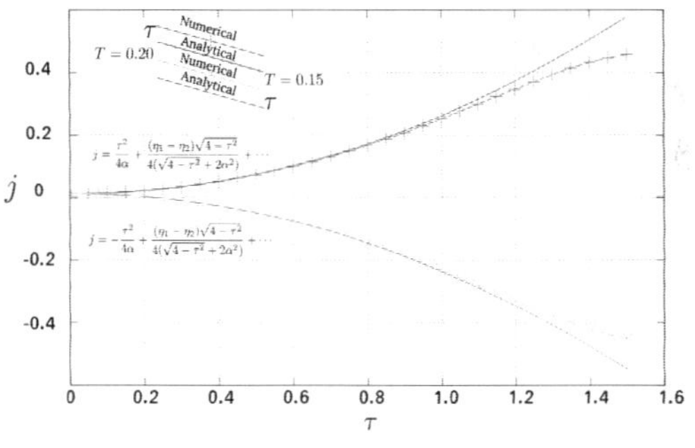

Figure 9. The energy flux when mechanical and thermal driving co-existed, $\alpha = 1.0$.

Fokker-Planck equation up to the first order gaves the overall local energy flux in synchronized rotation state[26], see FIG. 9.

$$\begin{aligned} j &= j_w + j_h, \\ &= \frac{\tau^2}{4\alpha} + \frac{(\eta_1 - \eta_2)\sqrt{4-\tau^2}}{4(\sqrt{4-\tau^2} + 2\alpha^2)} + \cdots . \end{aligned} \tag{13}$$

Comparing (12) and (13) one could identify the overall energy flux is seperated into the contributions of mechanical term and thermal term. More importantly, as the thermal term suggested, the mechanical and thermal contribution are not independent. More specifically, mechanical torque hampers the heat flux.

When the system is in the unsynchronized rotation state, the interaction between thermal and mechanical driving could be studied by numerical simulation.

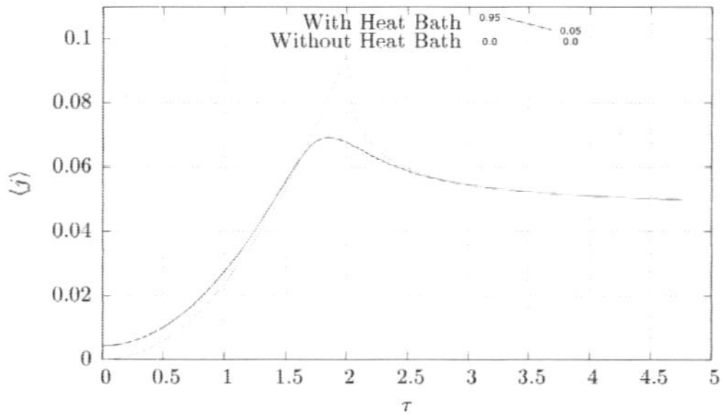

Figure 10. The comparison between a coupled rotator with and without a heat baths, $\alpha = 10.0$

Even with a high temperature difference between the heat baths, it is clear that the two curves of energy flux are qualitatively identical in the unsynchronized rotation state, which implies the contribution of heat baths is completely suppressed in the unsynchronized rotation state. Heat baths also broadens the transition point $\tau = 2$ to a region, in which the energy flux is reduced compare to its deterministic counterpart. This result suggests that in the transition region, the coupled rotator constantly switches between synchronized rotation and unsynchronized rotation. Thus we conclude that the reduction of local energy flux is due to the heat baths induced transition between different dynamical states.

Thus by studying a reduced model of coupled rotator lattice, we found that in the synchronized rotation state, mechanical and thermal contributions to the overall local energy flux was synergistic and in the unsynchronized (non-linear rotation modes) rotation state the contribution of heat baths is suppressed . More importantly, the reduction of heat flux is the consequence of transition between dynamical states. These results provide strong evidences that the relative rotation hampers the energy conduction, thus give an explanation to the finite heat conductivity of coupled rotator lattice.

4. Conclusion

In this paper, we have shown that the non-linear rotation mode on the coupled rotator lattice could lead to single peaked temperature profile and reduction of energy flux.

Understanding the mechanism of the energy flux reduction could lead to an understanding of the reason behind the normal heat conduction on coupled rotator lattice. It would also be interesting to explore the scaling of the energy flux plateau and the interface region in the clustered synchronous rotation state to obtain more knowledge about the non-linear rotational modes. Finally, one should take the non-linear rotational modes into consideration when investigating the heat conduction problems whenever the interaction potential permits such modes exist.

Acknowledgement

This work has been financially supported by grants from the National Natural Science Foundation of China (11475022).

References

[1] Peierls R E 2001 *Quantum Theory of Solids (Oxford Classic Texts in the Physical Sciences)* (Oxford University Press)
[2] Hu B, Li B and Zhao H 1998 *Phys. Rev. E* **57** 2992
[3] Lepri S, Livi R and Politi A 1997 *Phys. Rev. Lett.* **78**(10) 1896
[4] Maeda A and Munakata T 1995 *Phys. Rev. E* **52**(1) 234
[5] Narayan O and Ramaswamy S 2002 *Phys. Rev. Lett.* **89**(20) 200601
[6] Pereverzev A 2003 *Phys. Rev. E* **68** 056124
[7] Prosen and Campbell 2000 *Phys. Rev. Lett.* **84** 2857–2860
[8] Casher A and Lebowitz J L 1971 *J. Math. Phys.* **12** 1701
[9] Rieder Z, Lebowitz J L and Lieb E 1967 *J. Math. Phys.* **8** 1073
[10] Dhar A 2008 *Advances in Physics* **57** 457
[11] Lepri S, Livi R and Politi A 1997 *Phys. Rev. Lett.* **78** 1896
[12] Li B, Wang J, Wang L and Zhang G 2005 *Chaos* **15** 015121
[13] Li N and Li B 2007 *EPL* **78** 34001
[14] Piazza F and Lepri S 2009 *Phys. Rev. B* **79** 094306
[15] Wang J S and Li B 2004 *Phys. Rev. Lett.* **92** 074302
[16] Wang L, Hu B and Li B 2013 *Phys. Rev. E* **88**(5) 052112
[17] Fillipov A, Hu B, Li B and Zeltser A 1998 *Journal of Physics A: Mathematical and General* **31** 7719
[18] Casati G, Ford J, Vivaldi F and Visscher W M 1984 *Phys. Rev. Lett.* **52**(21) 1861
[19] Savin A V and Gendelman O V 2003 *Phys. Rev. E* **67**(4) 041205
[20] Nianbei Li Peiqing Tong B L 2006 *EPL* **75** 49
[21] Giardinà C, Livi R, Politi A and Vassalli M 2000 *Phys. Rev. Lett.* **84**(10) 2144
[22] Gendelman O V and Savin A V 2000 *Phys. Rev. Lett.* **84**(11) 2381
[23] Flach S and Gorbach A V 2008 *Physics Reports* **467** 1
[24] Iacobucci A, Legoll F, Olla S and Stoltz G 2011 *Phys. Rev. E* **84**(6) 061108
[25] Ke P and Zheng Z G 2014 *Frontiers of Physics* **9** 511
[26] Ke P 2014 Ph.D. Thesis. Beijing Normal University

Generating function technique in complex networks

Ming Li and Bing-Hong Wang*

Department of Modern Physics, University of Science and Technology of China, Hefei 230026, People's Republic of China

E-mail: *bhwang@ustc.edu.cn

Abstract. Complex network is a useful tool in describing the interactions between different agents of a complex system, and has attracted considerable interest recently. There are various approaches to studying the structure and dynamic of complex networks for the diverse and complicated connectivity structure. One more applicable theoretical approach is the generating function technique, which is useful for the studies of network structure and dynamic, especially for tree-like networks. In this paper, we will give a summary of the basic ideas of this approach, and explore the structure of networks. A cascading failures model we have proposed before will also be presented as an application of this approach.

1. Introduction

Network is a broad concept which can come into various aspects of our lives, such as communication network, Internet, relationship network, neural network and metabolic network. Complex network theory is such a uniform research platform for the topology structure and dynamic of networks in different domains[1]. For example, epidemic disease among people, spread of rumours online and cascading failures on the transport network can all be studied by the network percolation model[2, 3, 4].

As an interdiscipline subject, the methods and models available for complex networks are varied. One of the important theoretical approaches is the generating function technique[5, 6, 7], which is based on the branching process. This approach is often used to study the structure of tree-like networks, and the propagating processes that can be mapped to a network percolation problem[6, 7], such as epidemic and cascading failures. In other words, the generating function technique is a mean field method for the percolation on tree-like random networks. In addition, this approach is concerned with a network ensemble with a given degree distribution, not a network with given connection. The results obtained by this technique are the average of the ensemble, which will be consistent with the simulation (or empirical) results on a special network with size $N \to \infty$.

In the paper, we will give a brief review and discussion of the generating function technique with some examples.

2. Method

2.1. generating function

Generating function sometimes refers to as z-transform is widely used in mathematics, and plays an important role in probability theory[5]. Considering a sequence $a_n, n = 0, 1, 2, ...$, the corresponding generating function is defined as

$$G(x) = \sum_{n=0} a_n x^n. \tag{1}$$

For many well-defined sequences, $G(x)$ can be written in a closed form, and the coefficient a_n in the sequence can be recovered by Taylor expand,

$$a_n = \frac{1}{n!}\frac{d^n G(x)}{dx^n}\bigg|_{x=0}. \tag{2}$$

Considering a discrete random variable A, taking non-negative integer values. If a_n in eq.(1) is the probability that A takes the value $n(= 0, 1, 2, ...)$, $G(x)$ is also called as probability generating function. In this way, the expected value of A can be expressed as $G'(1)$. In addition, a probability generating function $G(x)$ must satisfy $G(1) = 1$. Taking Poisson distribution $a_n = \lambda^n e^{-\lambda}/n!$ as an example, the probability generating function is $G(x) = e^{\lambda(x-1)}$. Obviously, we have $G(1) = e^{\lambda(x-1)}|_{x=1} = 1$ and $G'(1) = \lambda e^{\lambda(x-1)}|_{x=1} = \lambda$.

2.2. generating function in complex networks

Next, we turn our attention to complex networks. In a network, the focus is usually concentrated on the states of nodes, which generally depend on the states of the links connected to them. For example, if all the links of a node are not connected to the giant component of the network, the node must not belong to the giant component.

For convenience, let x be the probability that a link is in state X, where X is relevant to the problem we considered. Further, we hypothesize that the states of links are independent of each other. Then, the probability that all the links of a node with degree k are in state X is x^k. Averaging this probability over the degree distribution of the network, we can obtain the fraction of the nodes with all their links in state X, in other words, the probability that all the links of a randomly chosen node are in state X. According to the definition of the generating function (1), this fraction (or probability) can be written as

$$G_0(x) = \sum_{k=0}^{\infty} p_k x^k. \tag{3}$$

Let $x = 1$, thus $G_0(1) = \sum_{k=0}^{\infty} p_k = 1$. This can be understood as that when the state of link is sure to X, this fraction (or probability) must be 1. Furthermore, the average degree of the network can be expressed as

$$G_0'(1) = \sum_{k=0}^{\infty} p_k k = \langle k \rangle. \tag{4}$$

In this way, the average degree could be understood as the rate of change of the node state with link.

More generally, for a node with degree k, the probability that there are only $i(\leq k)$ links are in state X can be written as

$$\binom{k}{i} x^i (1-x)^{k-i}. \tag{5}$$

Using the degree distribution of the network, the probability that a randomly chosen node are in this state (or the fraction of nodes in this state) can be written as

$$\sum_{k=i}^{\infty} p_k \binom{k}{i} x^i (1-x)^{k-i} = \frac{(-x)^i}{i!}\frac{d^i}{dx^i} G_0(1-x). \tag{6}$$

This function is more general than eq.(3) to represent the state of a node. However, it must be pointed that eq.(5) can reduce to x^k by letting $i = k$, but eq.(6) can not reduce to eq.(3).

Functions (3) and (6) describe the state of a randomly chosen node in a network. However, the nodes considered in practical network problems are not always chosen randomly, but rather ones reached by following randomly chosen links. So, we need to further explore the state of such a node.

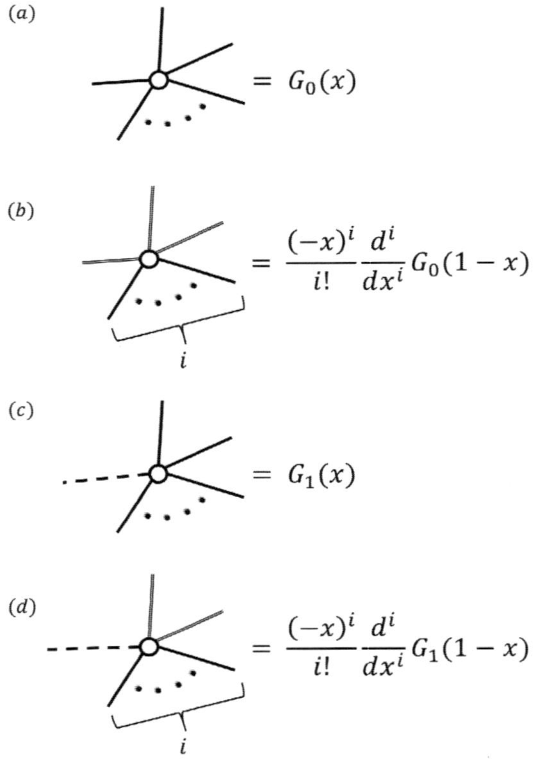

Figure 1. The generating functions for node states. Each link in the network is in state X with probability x. In the figure, links of state X are represented by solid lines, and empty lines for those not in state X. (a) All links of a randomly chosen node are in state X. (b) Only i links of a randomly chosen node are in state X. (c) All the other links of a node reached by following a randomly chosen link are in state X. (d) Only i links in the other links of a node reached by following a randomly chosen link are in state X.

Supposing the degree of the node reached by following a randomly chosen link is k, then, the probability that all the other links of this node are in state X is x^{k-1}. This probability can be used to describe the state of this node. Note that the link used for reaching this node must be ruled out. To obtain the average probability as eq.(3), we need to know the distribution of the node degree reached by following a randomly chosen link, the so-called excess degree distribution q_k. Obviously, q_k depends on the degree distribution p_k, the larger p_k is, the larger q_k is. In addition, following a link, it is easier to reach a node with larger k. In consideration of the above two points, we have

$$q_k \propto p_k k. \tag{7}$$

The normalized form of q_k is

$$q_k = \frac{p_k k}{\sum_k p_k k} = \frac{p_k k}{\langle k \rangle}. \tag{8}$$

Therefore, the average probability that all the other links of a node reached by following a randomly chosen link are in state X is

$$G_1(x) = \sum_{k=1}^{\infty} q_k x^{k-1} = \sum_{k=1}^{\infty} \frac{p_k k}{\langle k \rangle} x^{k-1}. \tag{9}$$

Note that the smallest degree of the node reached by following a link is 1. Obviously, eq.(9) satisfies $G_1(1) = 1$ and $G_1(x) = G_0'(x)/G_0'(1)$.

Furthermore, similar with eq.(6), the average probability that only $i(\leq k-1)$ links in the other links of a node reached by following a randomly chosen link are in state X, can be written as

$$\sum_{k=i+1}^{\infty} q_k \binom{k-1}{i} x^i (1-x)^{k-1-i} = \frac{(-x)^i}{i!} \frac{d^i}{dx^i} G_1(1-x). \tag{10}$$

Now that we have represented the common node states in a network shown graphically in fig.1. It must be pointed that all these functions are for the case that the states of links are independent of each other. For networks with correlated links, these functions do not apply. However, an approximate expression can be obtained by revising the functions shown in fig.1 for some special cases[8, 9, 10], this point is not considered here.

3. Network structure

3.1. hierarchical structure of networks

Hierarchical structure is one of the important characters of real networks[11]. In this section, we will give a brief discussion on the hierarchical structure of networks using generating function.

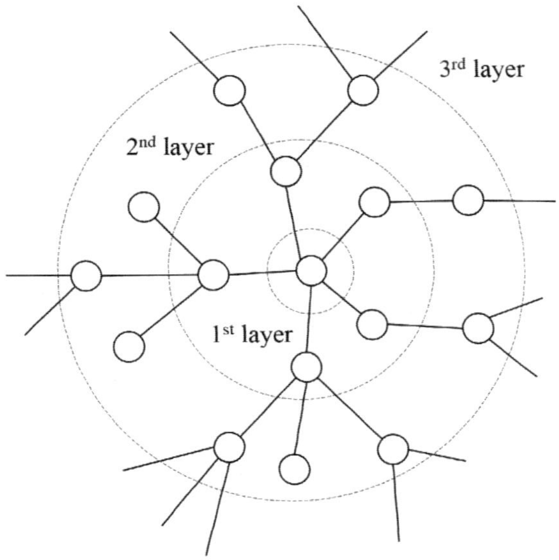

Figure 2. Hierarchical structure of networks. The number of nodes in the $(m+1)$-th layer is equal to the total excess degrees of the nodes in the m-th layer.

As shown in fig.2, we choose a node randomly as the center of the network, and then the direct neighbors of this node are the first layer of the network, and the neighbors' neighbors are the second layer, and so on. Since the central node is chosen randomly, its state can be described by the generating function $G_0(x)$. Here, state X means the connection of links. Following the links of the central node, we will get the first layer of the network. As mentioned last section, the generating function $G_1(x)$ can represent the state of the node reached by following a link. So, replacing x in $G_0(x)$ with $G_1(x)$, we will obtain the state of the first layer $G^{(1)}(x) = G_0(G_1(x))$. Similarly, the state of the second layer can be represented as $G^{(2)}(x) = G_0(G_1(G_1(x)))$. Generally, the state of the m-th layer can be written as

$$\begin{aligned} & G^{(m)}(x) \\ = \; & G_0(\underbrace{G_1(G_1(...G_1(x)...))}_{m}) \qquad (11) \\ = \; & G^{(m-1)}(G_1(x)), \quad m > 1. \qquad (12) \end{aligned}$$

Note that the state of links must be independent of each other, i.e., loop is forbidden.

For the state function $G^{(m)}(x)$ of the m-th layer, the integer power exponent of x must be the number of links that connect to the nodes in the $(m+1)$-th layer, since x represents the states of the excess links of the m-th layer. Therefore, the derivative of eq.(12) with respect to x at $x = 1$ must equal the total excess degrees of the nodes in the m-th layer, that is, the number of nodes in the $(m+1)$-th layer (see fig.2),

$$\begin{aligned} \langle k_{m+1} \rangle &= \left. \frac{dG^{(m)}(x)}{dx} \right|_{x=1} \qquad (13) \\ &= \left. \frac{dG^{(m-1)}(x)}{dx} \right|_{x=1} G_1'(1) \qquad (14) \\ &= \langle k_m \rangle G_1'(1), \quad m \geq 1. \qquad (15) \end{aligned}$$

From eq.(15), we can also find that

$$\frac{\langle k_{m+1} \rangle}{\langle k_m \rangle} = G_1'(1), \quad m \geq 1. \qquad (16)$$

This indicates that the growth rate of the node number of the m-th layer with m is a constant $G_1'(1)$. For an infinite network, there must be an infinite number of layers, so the node number can not decrease with m, i.e.,

$$G_1'(1) = \frac{\langle k^2 \rangle - \langle k \rangle}{\langle k \rangle} \geq 1. \qquad (17)$$

This is the existence condition of the giant component in a tree-like network[6, 12].

For a network with finite size, the average number of layers $\bar{l}$ must be a finite number. Next, we will show how to obtain $\bar{l}$. Using eq.(16), the node number of the m-th layer can be written as

$$\begin{aligned} \langle k_m \rangle &= \langle k_1 \rangle \left(\frac{\langle k_2 \rangle}{\langle k_1 \rangle} \right)^{m-1} \qquad (18) \\ &= \langle k \rangle \left(\frac{\langle k^2 \rangle - \langle k \rangle}{\langle k \rangle} \right)^{m-1}, \quad m \geq 1. \qquad (19) \end{aligned}$$

Summing over all the layers, we will obtain the network size

$$N = 1 + \sum_{m=1}^{\bar{l}} \langle k_m \rangle. \qquad (20)$$

This yields

$$\bar{l} = \frac{\ln\left[(N-1)(\langle k^2 \rangle - 2\langle k \rangle) + \langle k \rangle^2\right] - 2\ln\langle k \rangle}{\ln(\langle k^2 \rangle - \langle k \rangle) - \ln\langle k \rangle}. \qquad (21)$$

For a real network, $N \gg \langle k \rangle$ and $\langle k^2 \rangle \gg \langle k \rangle$, thus $\bar{l}$ can be approximately represented as

$$\bar{l} = \frac{\ln N - \ln\langle k \rangle}{\ln\langle k^2 \rangle - \ln\langle k \rangle} + 1. \tag{22}$$

This result can be seen as an approximation of the average distance of a real network[13, 14], which indicates that $\bar{l} \sim \ln N$.

3.2. k-shell structure

As we pointed above, the hierarchical structure shown in fig.2 has an arbitrary center. In complex network, there is another hierarchical structure with a determined center, i.e., k-shell structure[15]. For a given network, the k-shell structure is determined.

To illustrate k-shell structure of networks, we need to take a moment here to give the definition of k-core of network. k-core is a subnetwork of a network, in which nodes are connected to one another by at least k paths. k-core can be obtained by removing repeatedly all the nodes of degree less than k. The result, if it exists, is the k-core of the network (see fig.3). It is easy to know that the size of $(k+1)$-core S_{k+1} is smaller than that of k-core S_k. The nodes that belong to k-core and not belong to $(k+1)$-core are called as k-shell of the network.

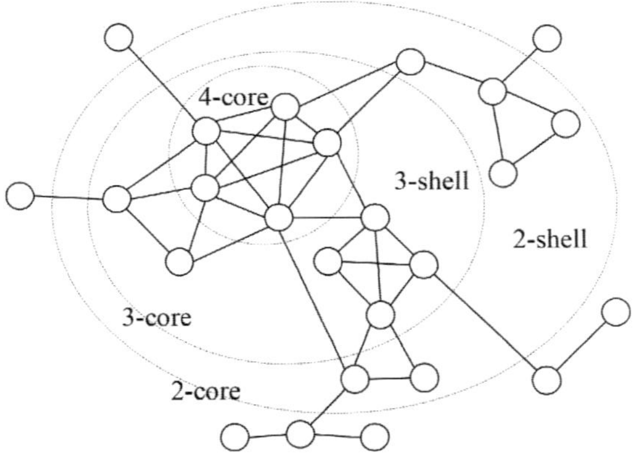

Figure 3. k-core and k-shell structure of networks. In k-core, nodes are connected to one another by at least k paths. The nodes that belong to k-core and not belong to $(k+1)$-core are called as k-shell of the network.

Next, we will give the size of k-core and k-shell of a network ensemble with given degree distribution. It must be pointed that k-core and k-shell can not exist in a finite tree-like network, so all the networks discussed here are infinite. For a node in k-core, it must have at less k links that connect to other nodes in k-core. Therefore, using the generating functions (6) and (10), we can get the size of k-core (or the fraction of nodes in k-core) directly,

$$S_k = \sum_{n=k} \frac{(-R_k)^n}{n!} \frac{d^n G_0(1-R_k)}{dR_k^n} \tag{23}$$

$$= 1 - \sum_{n=0}^{k-1} \frac{(-R_k)^n}{n!} \frac{d^n G_0(1-R_k)}{dR_k^n}, \tag{24}$$

$$R_k = \sum_{n=k-1} \frac{(-R_k)^n}{n!} \frac{d^n G_1(1-R_k)}{dR_k^n} \tag{25}$$

$$= 1 - \sum_{n=0}^{k-2} \frac{(-R_k)^n}{n!} \frac{d^n G_1(1-R_k)}{dR_k^n}. \tag{26}$$

Here, R_k means the probability that the node reached by following a randomly chosen link belongs to k-core. Note that the sum $\sum_n$ in eq.(25) begins with $k-1$, since the link used to reach the node has already provided a link for k-core. Solving the two equations, we can get the size of k-core S_k. Thus, the size of k-shell can be obtained,

$$L_k = S_k - S_{k+1}. \tag{27}$$

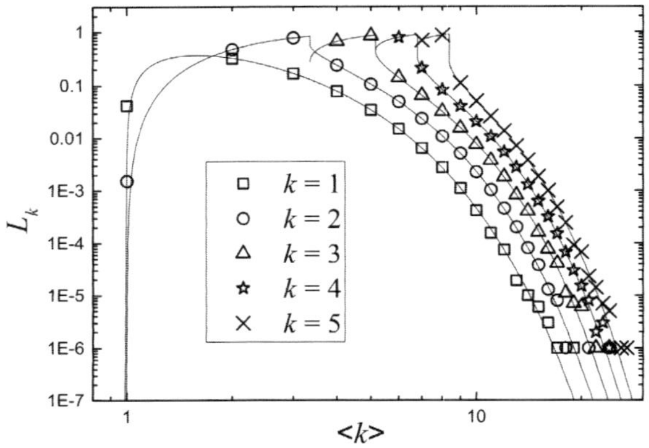

Figure 4. The size of k-shell of ER networks for different average degrees. In the simulation, the size of network is set as 10000. The solid lines are the corresponding theoretical predictions of eqs.(28) and (29).

For Erdős-Rényi (ER) network[12], the degree distribution is a Poisson distribution $p_k = e^{-\langle k\rangle}\langle k\rangle^k/k!$, thus eqs.(24) and (26) can be rewritten as

$$S_k = 1 - \Gamma(k, \langle k\rangle R_k), \tag{28}$$
$$R_k = 1 - \Gamma(k-1, \langle k\rangle R_k). \tag{29}$$

Here, $\Gamma(n, x)$ is the regularized incomplete gamma function. Figure 4 compares these theoretical results with the simulations. One can find the simulation results agree well with this theoretical analysis. When the average degree of the network is small, it is clear the size of k-shell is small. When $\langle k\rangle < 1$, there is no giant component in the network, so k-shell is obviously inexistent. With the increasing of the average degree of the network, k-shell with larger k will grow up and replace that of small k. Therefore, with the increase of the average degree, k-shell increases first and then decreases (see fig.4).

4. Robustness of networks

The robustness of networks is one of the important topics in complex networks[4, 16], which refers to the ability of a network to resist change without adapting its initial stable configuration. The typical robustness evaluation method is to consider the largest connected component in a network after a fraction $1-p$ of nodes are removed, the so-called giant component. The birth point p_c of a non-zero size of giant component S is often used to evaluate the robustness of the network. A small birth point p_c means the network is robust and resistant to the removal of a large number of nodes.

When we consider the robustness of networks, there are two link states in the network, i.e., connecting to the giant component or not. Let R be the probability that a link connects to the giant component. Thus, substituting $x = 1 - R$ into eqs.(3) and (9), we can obtain the probability that a randomly chosen node is not in the giant component $G_0(1-R)$, and the probability that a randomly chosen link does not connect to the giant component $G_1(1-R)$.

Next, we consider the cascading failures model proposed in ref.[17], which focuses on the robustness of networks with overlapping of connectivity and dependence links. Here, the connectivity link just is the ordinary link of network, and the dependence link represents the dependence of two nodes in a network. Specifically, a dependence link between node i and node j means that if node i fails, node j also fails, and vice versa. We consider a network connected by connectivity links with a degree distribution p_k and each node has exactly one dependence link. For a network with size N, there are $N/2$ dependence links. To represent the relation of the connectivity links and the dependence links, we assume that a fraction β of the dependence links overlaps with the connectivity links, and the other dependence links are set randomly. That means a fraction β of nodes will be adjacent to their dependence partners connected by the dependence links.

The cascading failure process begins with removing a fraction $1-p$ of nodes randomly. We assume that all the nodes not belong to the giant component will fail. As the model setting, the node with a failed dependence partner will also fail. So, two dependence nodes can be preserved after the cascading failures, only when they belong to the giant component simultaneously.

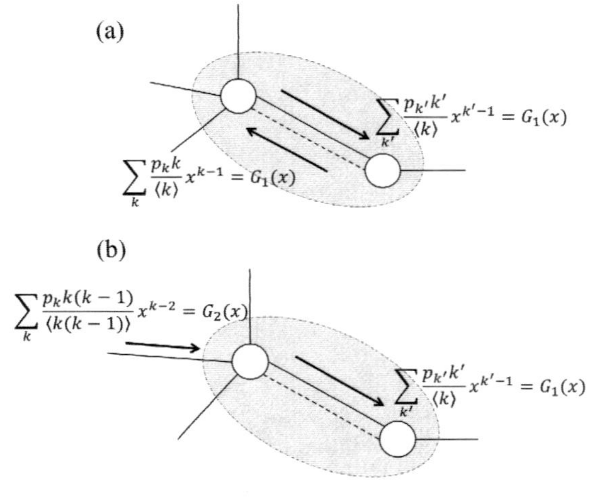

Figure 5. The generating functions for the states of the adjacent dependence nodes. The solid and dashed lines represent the connectivity and dependence links, respectively. (a) A dependence pair that chosen randomly from all the dependence pairs. The state of this pair can be described by generating function $[G_1(x)]^2$. (b) A dependence pair that reached by following a randomly chosen link. The state of this pair can be described by generating function $G_2(x)G_1(x)$, where $G_2(x) = G_0''(x)/G_0''(1)$.

To obtain the size of the giant component S, we need the generating function to express the state of a dependence pair, instead of that of a node. For nonadjacent dependence nodes, the dependence partner is a randomly chosen one, whose state can be described by generating function $G_0(x)$. So, the states of nonadjacent dependence pairs can be described by the joint state of two nodes. As shown in fig.5, the states for the adjacent dependence pairs can be described by generating functions $[G_1(x)]^2$ and $G_2(x)G_1(x)$, where $G_2(x) = G_0''(x)/G_0''(1)$.

Now, let $x = 1 - R$, we can obtain the equations for S and R,

$$S = (1-\beta)\, p^2 \left[1 - G_0(1-R)\right]^2 + \beta p^2 \left\{1 - \left[G_1(1-R)\right]^2\right\}, \quad (30)$$

$$R = (1-\beta)\, p^2 \left[1 - G_1(1-R)\right]\left[1 - G_0(1-R)\right] + \beta p^2 \left[1 - G_2(1-R)G_1(1-R)\right]. \quad (31)$$

For an arbitrary degree distribution p_k and the fraction of initial removal $1-p$, we can solve eq.(31) to obtain R, and then insert it into eq.(30) to obtain the order parameter S. In fig.6, we give the simulation results for ER networks, which agree with the analytical results well. In addition, we can find that there are two types of phase transition in this model, which can also be predicted by eq.(31) (see ref.[17] for details). For ER networks, we can also find that the tricritical point is a constant $\beta_c = \frac{1}{3}$.

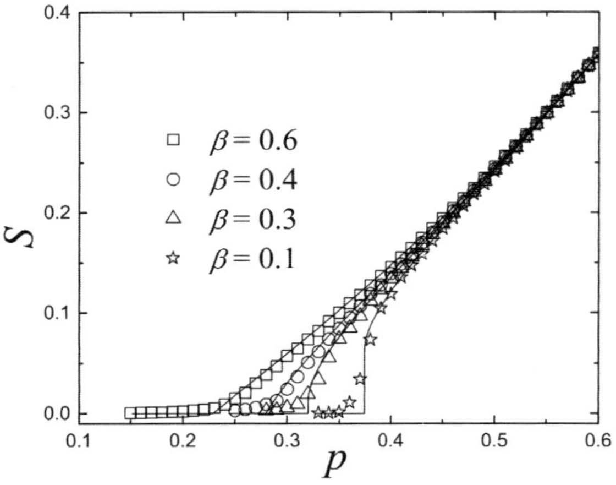

Figure 6. The size of the giant component versus the fraction of nodes that have been left after random removal for ER networks. The average degree of the network is $\langle k \rangle = 16$ and the network size is 10000. The solid lines are the corresponding theoretical predictions of eqs.(30) and (31).

5. Conclusion

In this paper, we have summarized the generating function technique used in the studying of complex networks, and explored the hierarchical and k-shell structures of networks using this technique. A cascading dynamic model has also been studied in this paper as an example that how to use this technique to solve the network dynamic model. In addition to the topics covered in this paper, there are a number of problems can be solved by the generating function technique, such as spread of epidemic disease on networks[7], observability transition[18], catastrophic cascade of failures in interdependent networks[19], clique percolation[20]. In the other hand, as mentioned in this paper, this technique can only give an exact result for tree-like random networks, and further research is needed to clustering and correlated networks.

Acknowledgments

This work is funded by the National Natural Science Foundation of China (Grant Nos.: 11275186 and 91024026) and the Open Funding Programme of Joint Laboratory of Flight Vehicle Ocean-based Measurement and Control (Grant No.: FOM2014OF001).

References

[1] Newman M E J 2010 *Networks: An introduction* (New York: Oxford University Press)
[2] Albert R and Barabási A L 2002 Statistical mechanics of complex networks *Rev. Mod. Phys.* **74** 47
[3] Dorogovtsev S N, Goltsev A V and Mendes J F F 2008 Critical phenomena in complex networks *Rev. Mod. Phys.* **80** 1275
[4] Cohen R and Havlin S 2010 *Complex networks: Structure, robustness and function* (Cambridge: Cambridge University Press)
[5] Wilf H S 1994 *Generatingfunctionology* (London: Academic Press)
[6] Newman M E J, Strogatz S H and Watts D J 2001 Random graphs with arbitrary degree distributions and their applications *Phys. Rev.* E **64** 026118
[7] Newman M E J 2002 Spread of epidemic disease on networks *Phys. Rev.* E **66** 016128
[8] Newman M E J 2003 Ego-centered networks and the ripple effect *Soc. Netw.* **25** 83
[9] Serrano M A and Boguñá M 2006 Clustering in complex networks: II. Percolation properties *Phys. Rev.* E **74** 056115
[10] Berchenko Y, Artzy-Randrup Y, Teicher M and Stone L 2009 Emergence and size of the giant component in clustered random graphs with a given degree distribution *Phys. Rev. Lett.* **102** 138701
[11] Shao J, Buldyrev S V, Braunstein L A, Havlin S and Stanley H E 2009 Structure of shells in complex networks. *Phys. Rev.* E 2009 **80** 036105
[12] Bollobás B 2001 *Random graphs* (Cambridge: Cambridge University Press)

[13] Newman M E J 2001 Scientific collaboration networks: I. Network construction and fundamental results *Phys. Rev.* E **64** 016131
[14] Newman M E J 2001 Scientific collaboration networks: II. Shortest paths, weighted networks, and centrality *Phys. Rev.* E **64** 016132
[15] Dorogovtsev S N, Goltsev A V and Mendes J F F 2006 k-core organization of complex networks *Phys. Rev. Lett.* **96** 040601
[16] Albert R, Jeong H and Barabási A L 2000 Error and attack tolerance of complex networks *Nature* **406** 378
[17] Li M, Liu R-R, Jia C-X and Wang B-H 2013 Critical effects of overlapping of connectivity and dependency links on percolation of networks *New J. Phys.* **15** 093013
[18] Yang Y, Wang J and Motter A E 2012 Network observability transitions *Phys. Rev. Lett.* **109** 258701
[19] Buldyrev S V, Parshani R, Paul G, Stanley H E and Havlin S 2010 Catastrophic cascade of failures in interdependent networks *Nature* **464** 1025
[20] Palla G, Derényi I and Vicsek T 2007 The critical point of k-clique percolationin in the Erdős-Rényi graph *J. Stat. Phys.* **128** 219

SPMCS2014 IOP Publishing
Journal of Physics: Conference Series **604** (2015) 012014 doi:10.1088/1742-6596/604/1/012014

Risk assessment and stock market volatility in the Eurozone: 1986-2014

Rui Menezes[a] and Álvaro Oliveira[a]

[a] ISCTE-IUL and BRU-IUL

E-mail: rui.menezes@iscte.pt

Abstract. This paper studies the stock market return's volatility in the Eurozone as an input for evaluating the market risk. Stock market returns are endogenously determined by long-term interest rate changes and so is the return's conditional variance. The conditional variance is the time-dependent variance of the underlying variable. In other words, it is the variance of the returns measured at each moment *t*, so it changes through time depending on the specific market structure at each time observation. Thus, a multivariate EGARCH model is proposed to capture the complex nature of this network. By network, in this context, we mean the chain of stock exchanges that co-move and interact in such a way that a shock in one of them propagates up to the other ones (contagion). Previous studies provide evidence that the Eurozone stock exchanges are deeply integrated. The results indicate that asymmetry and leverage effects exist along with fat tails and endogeneity. In-sample and out-of-sample forecasting tests provide clear evidence that the multivariate EGARCH model performs better than the univariate counterpart to predict the behavior of returns both before and after the 2008 crisis.

1 Introduction

Many studies on stock market volatility use a univariate approach and/or are mostly concerned with modeling the mean parameters of the underlying distribution function. However, understanding the variance effect and its evolution over time is also a major research topic in financial markets, since the variance can provide important information about the asset's risk and its evolution over time. Moreover, if we can model simultaneously the mean and the variance effects the approach becomes even more informative and it works in a fully (or quasi-fully) endogenous framework. Therefore, an investor can make more accurate decisions if she knows more deeply how risky is the investment she is about to make.

Higher risk is associated with larger price variations over time and conversely for lower risk. Higher risk is also associated with stock market price declines. This path can move due to changes in other market parameters or to the specific time moment of the measurement. In other words, volatility may depend not only on the behavior of other market variates but it may be also time dependent. Thus, an appropriate environment for studying market volatility may be the use of GARCH-type models. In the next section, we shall discuss these models in more detail. The third section describes the empirical data and results. The original data consist of stock market price indices and long-term interest rates for eleven Eurozone countries over the period 1986-2014. This means that the time window of our analysis covers the pre and the post 2008 crisis period. During the crisis period the long-term interest rates increased as well as market instability. How much was this effect? How long did it take to return to a near equilibrium situation if at all? Section 4 presents the major conclusions of this work.

2 Background and Methodology

2.1 Background

As the globalization process deepens and the economic agents compete in world-wide markets, the internationalization of the financial sources is also intensified. It makes a big difference to compete in a small market with 10 million inhabitants or another one with 500 million. When the price of commodities tends to uniformity across markets it makes sense that financial returns play a greater role in the financial management of firms or other economic agents. Economic agents will tend to search for the most profitable financial sources at the lowest market risk. However, there is a trade-off between financial returns and risk. Higher expected returns are usually associated with higher expected risk and the investor wishes to maximize the former and minimize the latter. The equilibrium point between returns and risk depends on the risk profile of each investor. Assuming that the returns are exogenously and *a priori* determined, the investor will look, basically, at the risk.

There are many definitions and dimensions of risk. There are also a large number of ways to measure risk. In this study, the volatility in the stock market represents the risk. It is a dispersion measure since its computation derives from the conditional variance obtained out of the estimation of a GARCH-type model. The conditional variance is a time series that represents the estimated variance level of the underlying asset at each moment in time. Low volatility markets are always preferable to high volatility markets, at least if the investor is risk-averse and not a speculator. In addition, the volatility increases when the market prices decrease and vice-versa. Therefore, higher stock market instability is linked with bear stock markets.

Across the whole sample period under analysis (1986-2014), there are two distinct subperiods. The first one ended in 2007/08 with the worldwide economic crisis. Bull stock markets characterize the most part of this relatively tranquil period, with patchy episodes of short-term cracks usually motivated by very specific events. However, none of these cracks had the dimension of the 2008 crisis. Beginning in the US, the global crisis of 2008 affected mainly Europe and in particular the Eurozone economies. It lasted up to 2014 in some cases. Markets turned out to be more turbulent and volatile. Interest rates increased dramatically in many countries due to the sovereign debt crisis that affected mainly the southern European economies. Stock market prices declined very sharply for a couple of years and the bailout-rescued countries (Greece, Portugal, and Ireland) were obviously among the hardest affected. The crisis persisted for a while but in most countries, the stock market started to recover again and volatility went back to lower levels. In this paper, we shall call pre-crisis sample to the period 1986-2008 and post-crisis sample to the period 2008-2014. The breakpoint occurred in September 2008 as was endogenously confirmed by the Lanne-Lütkepohl-Saikkonen [1] unit root test with structural breaks.[1]

2.2 Conditional Heteroskedasticity

In financial data applications, a popular model of volatility is the Generalized AutoRegressive Conditional Heteroskedasticity or GARCH(p, q). This model, formerly

[1] The results of these tests are available from the authors upon request.

proposed by [2], generalizes the original ARCH(q) model introduced by [3]. A simple version of the GARCH(p, q) model with k exogenous variables in the mean equation is:[2]

$$y_t = \mu_0 + \sum_{i=1}^{k} \mu_i x_{it} + \varepsilon_t \,, \tag{1}$$

$$\varepsilon_t = v_t \sigma_t \,, \tag{2}$$

$$\sigma_t^2 = \alpha_0 + \sum_{j=1}^{q} \alpha_j \varepsilon_{t-j}^2 + \sum_{i=1}^{p} \beta_i \sigma_{t-i}^2 \,, \tag{3}$$

where $\varepsilon_t \mid I_{t-1} \sim N(0, \sigma_t^2)$; $v_t \sim iid(0, 1)$; $\alpha_0 > 0$; $\alpha_1, \ldots, \alpha_q, \beta_1, \ldots, \beta_p \geq 0$ and $\alpha_1 + \ldots + \alpha_q + \beta_1 + \ldots + \beta_p < 1$. I_{t-1} denotes the information set available up to the moment t-1. The error term ε_t represents a random disturbance at time t and σ_t^2 represents its conditional variance. Equation (1) models the mean, that is, the $E(y_t \mid x_{it})$; equation (2) represents a factorization of the random disturbance into a innovation (v_t) and the conditional standard deviation of the error term (σ_t); equation (3) models the conditional variance. The conditional variance is a linear function of its own historical values up to lag p and the historical values of the squared residuals up to lag q.

The restriction $\alpha_1 + \ldots + \alpha_q + \beta_1 + \ldots + \beta_p < 1$ is required in order to assure the stability and covariance stationarity of ε_t.[3] If all the β parameters in (3) are simultaneously equal to zero then we have the ARCH(q) model. Notice that the GARCH(p, q) is asymptotically equivalent to the ARCH(∞) model and many conditional variance processes can be adequately represented by a GARCH(1, 1) model. In this case the unconditional variance of ε_t is:

$$\operatorname{var}(\varepsilon_t) = \sigma_\varepsilon^2 = \alpha_0 / (1 - \alpha_1 - \beta_1) \,, \tag{4}$$

with $\alpha_1 + \beta_1 < 1$.

In the GARCH(p, q) model the variance of the error term at the moment t, conditional on past errors is given by $\sigma_t^2 = \operatorname{var}(\varepsilon_t \mid \varepsilon_{t-j})$. σ_t^2 is an increasing function of the magnitude of past errors, independently of its sign, given that the variance equation uses squared errors. Therefore, positive or negative errors of large magnitude tend to follow

[2] Notice that p denotes the number of lags in the conditional variance and q denotes the number of lags in the squared error term.
[3] Otherwise, the regression equation (1) is spurious.

errors of the same kind and likewise for errors of small magnitude. This is known in the literature as volatility clustering and is a common feature of many financial time series based on returns.

Another important feature of many financial return's time series is leptokurtosis or fat tails. Leptokurtosis occurs when the shape of the distribution is more peaked than the bell-shaped Gaussian distribution. The distribution around the mean value is narrower but the tails are longer and fatter than that in the Gaussian distribution. This means that extreme events are more likely to happen (not so rare) and have a larger magnitude. Therefore, the assumption of random normal movements in the stock market returns (or price changes) does not seem to be adequate. The GARCH-type models allow us to use non-Gaussian distributions to describe the behavior of the errors ε_t. Among them, two popular distributions are the *t*-Student and the Generalized Error Distribution. Both are adequate to deal with leptokurtosis and volatility clustering. In this paper, we shall rely on the *t*-Student distribution of the error term.

Many studies on stock market volatility use GARCH-type models to infer about the time-path behavior of returns. Among them we find [4]-[17]. Other authors analyze the issue of volatility based on statistical physics concepts (see, *e.g.* [18] among others).

2.3 Asymmetric Volatility

The GARCH(p, q) model gives a symmetric treatment to the conditional variance, that is, positive and negative perturbations of the same magnitude have a similar effect over the estimated conditional variance. However, it is well known that such symmetry may not exist and that in many cases bad news and good news have different impacts on the behavior of volatility. Usually, the volatility is more sensible to bad news than to good news. A commonly cited example is the relationship between the price of oil and the price of gasoline. When the oil price increases the price of gasoline also increases. When the oil price decreases the price of gasoline decreases too but not with the same magnitude.

Nelson [19] suggested the use of an exponential GARCH model to deal with the issue of asymmetry. Nelson's model is known in the literature as the EGARCH(p, q). In addition to the issue of asymmetry, the EGARCH model offers the advantage of not imposing non-negativity restrictions on the variance equation parameters in order to secure the non-negativity of σ_t^2 over its sample domain. As for the GARCH, this model assumes short memory and null persistency. This means that the variables in the model are integrated of order zero [I(0)].

The variance equation of the EGARCH(p, q) model can be written as:[4]

$$\ln\left(\sigma_t^2\right)=\alpha_0+\sum_{j=1}^{q}\alpha_j\left(\frac{\left|\varepsilon_{t-j}\right|}{\sigma_{t-j}}-\sqrt{\frac{2}{\pi}}\right)+\sum_{k=1}^{m}\gamma_k\frac{\varepsilon_{t-k}}{\sigma_{t-k}}+\sum_{i=1}^{p}\beta_i\ln\left(\sigma_{t-i}^2\right). \tag{5}$$

The interpretation of α_j and β_i is similar to that of the GARCH(p, q) model. The main difference resides in the γ_k parameters. These parameters allow us to test both the presence of asymmetric effects and whether there are leverage effects. The effect is asymmetric if $\gamma_k \neq 0$. In addition, if $\gamma_k < 0$ there are leverage effects. In the EGARCH(p, q) model the leverage effects are exponential. Given that the variance equation is specified in order to the logarithm of the conditional variance (and not the conditional variance itself as in the GARCH model), the conditional variance forecasts are always positive.

We already know what the meaning of asymmetric volatility is: different magnitudes of the effect of similar positive and negative shocks in the market. Now, how can we interpret the leverage effect? The leverage effect occurs when the volatility increases more during bear market periods, and vice-versa, after a shock of similar magnitude. It can be seen as a particular case of asymmetry since it occurs when the relationship between returns and volatility is negative. (see, *e.g.* [20]-[22]).

Indeed, when the stock market share price declines and the firm's liabilities remain constant in the short-run, the firm's ratio debt/equity increases. The firm becomes more leveraged and its future more uncertain. Therefore, share prices will become more volatile. Additionally, when volatility increases more in result of negative rather than positive returns, the autocorrelation between squared returns in consecutive periods tends to be high and negative. This will cause asymmetric volatility clustering in the market.

Assuming in (5) that $p = q = 1$, the EGARCH(1,1) variance equation is given by:

[4] The mean equation is the same as in the GARCH model [Eq. (1)].

$$\ln\left(\sigma_t^2\right) = \alpha_0 + \alpha_1\left(\frac{\left|\varepsilon_{t-1}\right|}{\sigma_{t-1}} - \sqrt{\frac{2}{\pi}}\right) + \gamma_1\frac{\varepsilon_{t-1}}{\sigma_{t-1}} + \beta_1\ln\left(\sigma_{t-1}^2\right), \qquad (6)$$

where the interpretation of the parameters is similar to that of the EGARCH(p, q) model. Notice that in the above specification exogenous regressors only appear in the mean equation. The variance equation is fully endogenous and depends only on the residual's series of the mean equation and its estimated conditional variance. Thus, the robustness of the estimates of the variance parameters depends a lot on the correct specification of the mean equation. Yet, is it not possible that the conditional variance itself depends on exogenous regressors, along with endogenous ones? We shall explore this issue in the next subsection.

2.4 Multivariate Conditional Volatility

The asymmetric conditional heteroskedasticity model that we propose extends the original EGARCH(p, q) model to a multivariate framework. It includes a set of regressors in the variance equation in order to produce better estimates of the conditional variance time series. The parameters of the exogenous regressors in the variance equation are denoted by the vector $\theta = (\varphi_i, \delta_j)$. The estimation process of this modified EGARCH model poses no serious problems if the mean and variance equation regressors are independent of the error term ε_t and of its conditional variance σ_t^2. Otherwise, a Feasible Generalized Least Squares estimator can be constructed to estimate the model, such as proposed by [23].

The aim of this study is to analyze the behavior of the conditional variance of the stock market returns of some Eurozone countries over a relatively long time span. Such behavior may shift depending on interest rate changes and other economic and/or financial factors. The occurrence of a large crisis may change the structural behavior of stock market returns, at least while it lasts. Of course, many other economic and behavioral factors may affect the time-path of stock market returns. We shall keep our model as simple as possible and therefore we just model market returns as a function of long-term interest rate changes and a series of dummies denoting the post-crisis years. These dummies capture the specific effects of each year since the outset of the crisis (2008-2013). The augmented EGARCH(1, 1) that we propose takes the following form:

$$\ln\left(\sigma_t^2\right)=\alpha_0+\alpha_1\left(\frac{\left|\varepsilon_{t-1}\right|}{\sigma_{t-1}}-\sqrt{\frac{2}{\pi}}\right)+\gamma_1\frac{\varepsilon_{t-1}}{\sigma_{t-1}}+\beta_1\ln\left(\sigma_{t-1}^2\right)+\sum_{i=1}^{k}\varphi_i x_{it}+\sum_{j=1}^{m}\delta_j D_j\ , \qquad (7)$$

where the φ_i are the parameters of the continuous regressors and the δ_j are the parameters of the annual dummies. We only included one continuous regressor in (7): the long-run Treasury bill interest rate change (10 years). So x_{it} = $\Delta(\mathrm{TB}_t)$ where TB denotes the 10-year Treasury bill interest rate and Δ is the first difference operator. There are six dummies D_j in (7), one for each year over the time window 2008-2013. The mean equation is the same as shown in (1) where y_t denotes the stock market returns $[\ln(P_t/P_{t-1})]$ and x_{it} is as defined above. Therefore, the specification of our model is multivariate in the mean and in the variance and this is an important improvement relatively to previous studies that adopt a univariate framework. We proceed in the next section with the presentation of the data and main findings.

3 Data and Results

3.1 The Data

The data used in this paper consist of daily stock market price indices and 10-year Treasury bill interest rates for 11 countries that belong to the Eurozone. The sample spans from January 1st 1986 to January 3rd 2014, totalizing a maximum of 7305 data points for each time series.[5] The data points cover five days per week (Monday to Friday). All the countries in this study are included in the developed markets group. The data were collected in the Datastream database. All the daily stock market indices were converted to relative price indices where the base 100 corresponds to the first recorded observation. Stock market returns correspond to the first difference of log relative prices at each moment t. Interest rate changes are the first difference of observed interest rates at each moment t.[6]

---

[5] The series do not necessarily start at the same date for all the countries, thus the sample size may vary from series to series.
[6] Note that the first difference is given by $\Delta x_t = x_t - x_{t-1}$.

Figure 1 shows the time series of the daily stock market returns and 10-year interest rate changes for the 11 Eurozone countries over the sample period considered.

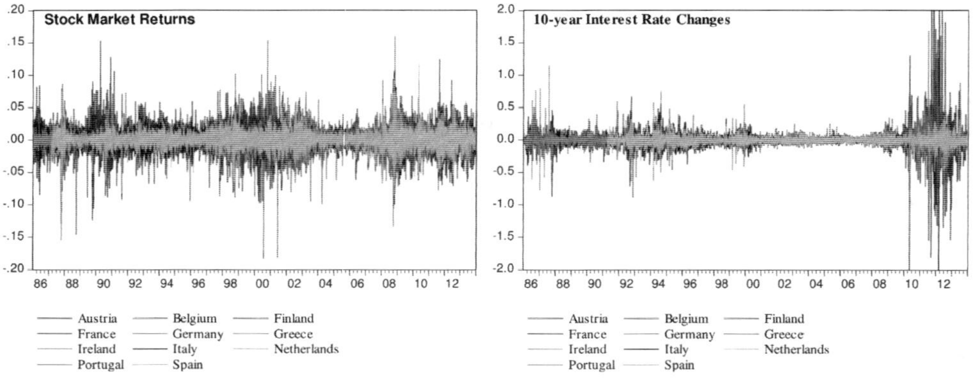

Figure 1. Daily stock market returns and 10-year interest rate changes for 11 Eurozone countries
Source: Datastream. Sample: 1/06/1986 to 1/03/2014.

The stock market returns panel shows two periods of high and persistent volatility: the first one approximately between 1997 and 2002, and the second one from the 2008 crisis onwards. The 10-year interest rate changes panel exhibits the interest rate instability period following the 2008 crisis with extreme peaks between 2010 and 2012. These extreme peaks correspond to the large rise in the long-term interest rates of the rescued countries. Towards the end of the sample period, the interest rates market becomes a bit more stable.

Table 1 presents some descriptive statistics of the stock market returns and 10-year interest rate changes over the sample period.

Some highlights of these results are: Greece (0.018) and Finland (0.017) have the higher unconditional standard deviation of the 11 stock market returns series analyzed. Greece (0.509) and to a lesser extent Portugal (0.095) have the higher standard deviation of the 10-year interest rate change time series analyzed. All means all close to zero. The Jarque-Bera test rejects the null of normality at the 1% level in all cases. This is consistent with previous findings that stock market returns (and also interest rate changes) should be modeled in a non-Gaussian framework. Volatility clustering and fat tails prevail as the main features of these variates.[7]

Table 1. Descriptive statistics: stock market returns and 10-year interest rate changes

Country	Stock Market Returns				10-year Interest Rate Changes			
	Mean	SD	J-B	*N*	Mean	SD	J-B	*N*
Austria	1.84E-04	0.011	29978	7305	-6.84E-04	0.040	7850	7164
Belgium	2.51E-04	0.010	32752	7305	-9.05E-04	0.047	9414	6414
Finland	2.50E-04	0.017	18870	6725	-1.51E-03	0.058	130525	5829
France	2.54E-04	0.012	10656	7305	-1.01E-03	0.055	64951	7284
Germany	1.72E-04	0.012	33911	7305	-6.01E-04	0.045	4773	7305
Greece	2.42E-04	0.018	10336	6784	5.84E-04	0.509	7.92E+08	3851
Ireland	2.90E-04	0.013	42020	7305	-8.25E-04	0.072	403673	7203
Italy	1.09E-04	0.013	6258	7305	-1.69E-03	0.071	47165	5957
Netherlands	1.97E-04	0.012	20925	7305	-6.16E-04	0.042	2903	7305
Portugal	6.50E-05	0.010	24744	6263	-9.10E-04	0.095	769748	5340
Spain	2.06E-04	0.013	9865	7004	-1.41E-03	0.064	43865	5936

Notes: J-B [24] normality test (H_0: the distribution of *X* is Gaussian. Rejection of the null at 1% in all cases). Sample: 1/06/1986 to 1/03/2014.

3.2 Multivariate EGARCH Results

Table 2 displays the parameter estimates of the mean and variance equations of the multivariate EGARCH model where the predictor variable is the daily 10-year interest rate change (first difference) and the predicted variable is the daily stock market returns (first difference of the log index). Both variables are I(0).

The mean equation is a static simple linear regression model that relates the predicted with the predictor variable. The variance equation is a dynamic model that relates the log conditional variance (volatility) with its historical values and the historical values of the mean equation residuals. Additionally, the log conditional variance also depends on the 10-year interest rate change and a set of six dummies for the post-crisis years. These variables are used as instruments in order to correctly isolate the true asymmetric effects of the stock market returns.

The interpretation of the mean equation parameter estimates is as usual. The slope coefficient μ_1 denotes the average impact of a unit variation in the interest rate change on stock market returns. All the estimates are significantly negative at the 1% level ex-

[7] Unit root tests were also carried out and rejected the null of nonstationarity at the 1% level. The results are available from the authors upon request.

cept the one for Finland. Thus, a positive shock in interest rates implies a decline in returns, which is consistent with theoretical expectations. The magnitude of the impact is larger for France and Italy (0.036) than for the remaining countries. μ_0 is the intercept and denotes the expected value of the returns when there is no change in the 10-year interest rate.

Table 2. Multivariate EGARCH parameter estimates: mean and variance equations

Country	Mean Equation				Variance Equation							
	μ_0		μ_1		α_0		α_1		γ_1		β_1	
Austria	0.001	**	-0.007	**	-0.543	**	0.247	**	-0.035	**	0.963	**
Belgium	0.000	**	-0.018	**	-0.395	**	0.200	**	-0.067	**	0.975	**
Finland	0.001	**	0.000		-0.151	**	0.121	**	-0.034	**	0.993	**
France	0.001	**	-0.036	**	-0.306	**	0.148	**	-0.083	**	0.979	**
Germany	0.001	**	-0.025	**	-0.288	**	0.155	**	-0.070	**	0.982	**
Greece	0.000		-0.005	**	-0.521	**	0.194	**	-0.056	**	0.958	**
Ireland	0.001	**	-0.014	**	-0.346	**	0.173	**	-0.037	**	0.977	**
Italy	0.000	*	-0.036	**	-0.317	**	0.173	**	-0.060	**	0.980	**
Netherlands	0.001	**	-0.011	**	-0.268	**	0.153	**	-0.070	**	0.984	**
Portugal	0.001	**	-0.008	**	-0.449	**	0.247	**	-0.050	**	0.973	**
Spain	0.001	**	-0.018	**	-0.289	**	0.143	**	-0.071	**	0.981	**

Notes: Method: ML - ARCH (Marquardt) - Student's t distribution. ** Significant at 1%. * Significant at 5%. φ_i and δ_j parameters are not reported but the estimates are available from the authors upon request. Sample: 1/06/1986 to 1/03/2014.

Regarding the variance equation we report the parameter estimates α_0, α_1, γ_1 and β_1. All these estimates are significant at the 1% level. α_0 denotes the intercept which is not constrained to be positive in the EGARCH specification. Note that a negative value of α_0 means that the expected conditional variance is lower than one when all the variates in the variance equation are equal to zero. α_1 is the coefficient of the ratio between the absolute value of the mean equation residuals and the conditional standard deviation at time t-1. γ_1 is the coefficient of the ratio between the residuals and the conditional standard deviation at time t-1. Finally, β_1 is the first-order autoregression coefficient of the log conditional variance (dependent variable).

As noted in section 2 the parameter γ_1 is the key to assess asymmetric and leverage effects in the volatility caused by past positive and negative shocks. There are asymmetric effects if $\gamma_1 \neq 0$. In addition, if $\gamma_1 < 0$ we observe exponential leverage effects. This occurs when bad news ($\varepsilon_{t-1} < 0$) imply an increase in volatility. Notice that $\varepsilon_{t-1} < 0$ happens when a negative shock at moment t-1 provokes a decrease of the underlying value of the predicted variable at moment t, relatively to the long-run equilibrium relationship. This is consistent with the previous findings that volatility (instability) increases when stock market prices decrease. A significant negative value of γ_1 under the EGARCH model implies that the larger is the absolute value of γ_1 the smaller is the leverage effect. This is the case of France (-0.083) and Spain (-0.071). Conversely, Finland (-0.034) and Austria (-0.035) exhibit the largest leverage effect over the whole sample period amongst the 11 countries analyzed.

3.3 In-sample Forecasts

Based on Eq. (7) one can obtain the estimates of the stock market return's conditional variance at each moment t. These estimates are the in-sample forecasts of the underlying volatility. Using two standard errors as boundaries one can construct confidence intervals around the static forecasts. When the interval of confidence is wider this means that the forecasted conditional variance (volatility) increases. Figure 2 displays the graphs of the in-sample volatility forecasts for the 11 countries analyzed.

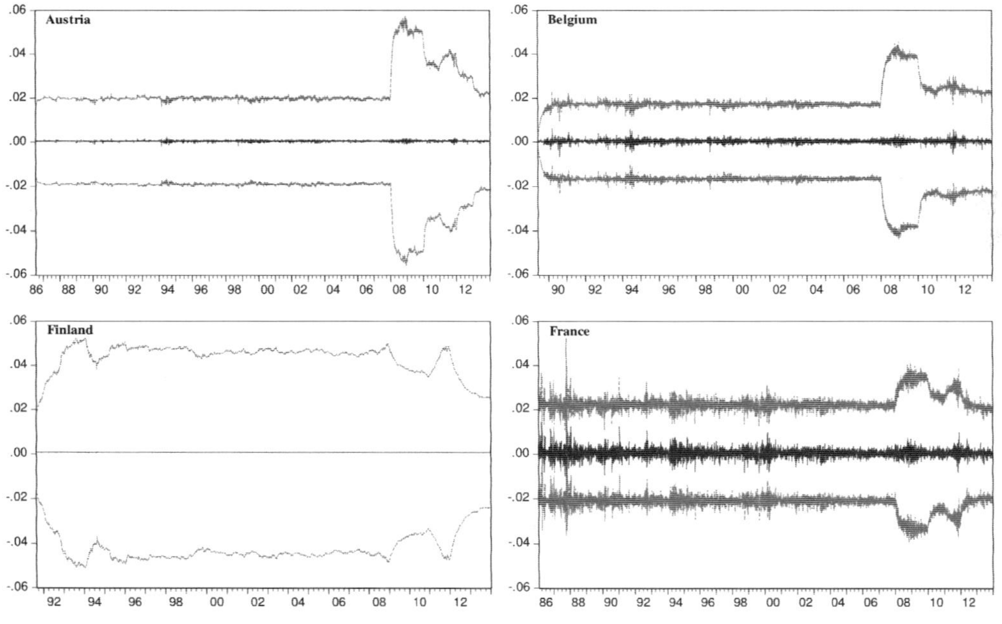

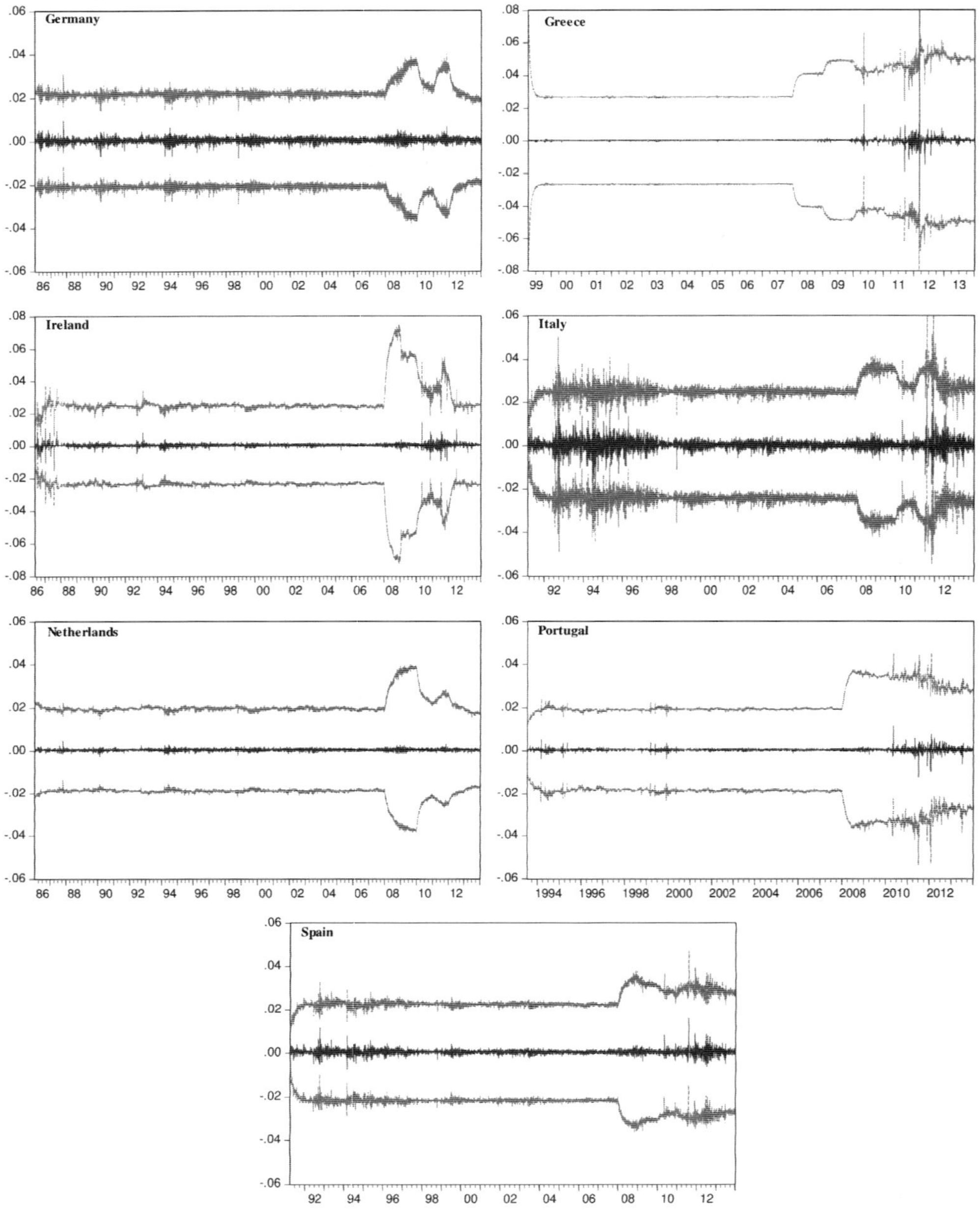

Figure 2. EGARCH volatility forecasts of returns for 11 Eurozone countries
Notes: Exogenous variable: 10-year interest rate changes. ---: ±2 S.E. Sample: 1/06/1986 to 1/03/2014.

The graphs depicted in Figure 2 are elucidative. In all cases except Finland, the boundaries of the confidence interval are close to -0.02 and 0.02 in the pre-crisis period. These boundaries increased dramatically during the crisis period and came back to the pre-crisis values towards the end of the sample window except in Greece, Portugal, and

Spain (the southern peripheral GPS countries). The peak of high volatility occurred in most cases in the last quarter of 2008 or 2009. A right-side tapered arrow indicates less persistence of the crisis and, therefore, a faster return to the pre-crisis volatility levels. In the case of the rescued countries, Ireland shows a quite fast recovery, which is not the case of Greece and Portugal, at least during the sample window analyzed. Finland exhibits high volatility since the 1990's and throughout the most part of the sample period.

To conclude, the EGARCH volatility forecasts appear to be consistent with the observed performance of the stock markets, in particular during the crisis and onwards. The distinction between the effect of good and bad news is remarkable and statistically significant. Likewise, the assumption of *t*-Student errors rather than Gaussian seems to be more realistic and robust given the leptokurtic feature of the return's data.

3.4 Out-of-sample Forecasts

The in-sample forecasts presented in the previous section show that the multivariate EGARCH model proposed in this paper is a good option to explain the time-path of the conditional variance even when market turbulence occurs due to exogenous factors. In our case, these exogenous factors include the 10-year interest rate change and the specific effects of the 2008 crisis and subsequent periods. However, the in-sample forecasts are not able to provide information about the forecasting accuracy of the model for future periods. Such information can be obtained from out-of-sample forecasting tests.

Out-of-sample forecasting tests allow us to select the model specification that yields better accuracy forecasts. To this end, we split the whole sample into the subsample used for estimation purposes and the subsample used for forecasting comparisons. We are especially interested in assessing whether the pre-crisis observations are useful to forecast the post-crisis ones. Thus, we re-estimate the models using just the pre-crisis observations. In order to obtain the post-crisis (out-of-sample) forecasts we use the estimated parameters. These forecasts can be compared with the observed values of the underlying forecasted variable. The more accurate forecasts we obtain the better and these will be used to select the best model specification.

The competing model specifications are the multivariate EGARCH and the univariate EGARCH (see, *e.g.* [25]). Although the Diebold-Mariano [26] forecasting accuracy test is a popular one, it has some drawbacks that prevent its use. In our case, as the forecasting horizon is long, the Diebold-Mariano test leads to problems of oversizing because the null hypothesis of equal forecasting accuracy tends to widen the confidence interval around H_0. An alternative is to test whether a set of forecasts encompasses a rival set ([27]-[28]). These tests allow us to compare the forecasting performance of manifold specifications although in this study we just compare two. We shall use the Harvey *et al.* [29] methodology extended later to allow for multiple comparisons by [30]. The results are presented in Table 3.

Table 3. Out-of-sample HLN forecasting tests

Country	*F*-standard		HLN	
	Test	*p*-value	Test	*p*-value
Austria	2.166	(0.151)	3.416	(0.066)
Belgium	3.253	(0.072)	1.769	(0.184)
Finland	1.981	(0.178)	3.528	(0.061)
France	2.531	(0.112)	3.390	(0.067)
Germany	2.578	(0.109)	3.243	(0.074)
Greece	22.073	(0.000)	13.816	(0.000)
Ireland	1.350	(0.246)	1.799	(0.180)
Italy	0.028	(0.866)	0.031	(0.861)
Netherlands	2.718	(0.100)	2.796	(0.095)
Portugal	4.885	(0.046)	3.252	(0.074)
Spain	0.128	(0.720)	0.115	(0.735)

Notes: H_0: the multivariate EGARCH encompasses the univariate EGARCH. H_0 is rejected if $p < 0.05$. Sample: 1/06/1986 to 1/03/2014.

The null hypothesis postulates that the multivariate EGARCH encompasses the alternative univariate EGARCH model. That is, the multivariate EGARCH has better forecasting accuracy than the univariate one. Having better forecasting accuracy means that the forecasting errors are smaller and we are minimizing the loss function. For each country, we present the statistics and the corresponding *p*-values (in brackets) for the *F*-standard test and the HLN test. The forecasting time series were obtained from one-step ahead static forecasts. The null is only rejected in the case of Greece. Italy and Spain present the most accurate multivariate EGARCH forecasts.

To summarize, volatility plays a central role in forecasting the stock market performance. Reliable forecasts are best obtained in a multivariate framework. This is especially the case when we use data observed over a tranquil period to predict the market behavior over a turbulent period. Therefore, volatility can be seen as a stock market risk measure. The higher is the volatility the bigger is the risk and the investor can either win or lose much more than in a period of lower volatility. The decision depends on the profile of the investor, that is, on whether she is risk-averse, risk-taker, or risk-neutral. For each profile, there is a ranking of stock markets to invest.

4 Conclusions

This paper analyses the stock market volatility of 11 Eurozone countries over the period 1986-2014. Volatility is modeled in a multivariate framework where interest rate changes and the economic crisis of 2008 help to explain part of the increase of market turbulence at that time. When interest rate changes are used to model the conditional variance of stock returns, the remaining part of the variance equation explains the endogenous behavior of volatility. While most markets adjusted relatively quickly to the pre-crisis levels of volatility, Greece, Portugal, and Spain did not return to their pre-crisis levels during the sample period analyzed. On the other hand, Finland presents high volatility levels at least since the outset of the 1990's. In-sample forecasts and out-of-sample forecasting tests show that the multivariate EGARCH model performs better than the univariate counterpart in all the cases except Greece. Our approach provides a better way to improve the forecasting ability of stock market returns.

References

[1] Lanne M, Lütkepohl H and Saikkonen P 2002 Comparison of unit root tests for time series with level shifts, *Journal of Time Series Analysis* **23(6)**, 667-685.

[2] Bollerslev T 1986 Generalized autoregressive conditional heteroskedasticity, *Journal of Econometrics* **31(3)**, 307-327.

[3] Engle R F 1982 Autoregressive conditional heteroscedasticity with estimates of the variance of United Kingdom inflation, *Econometrica* **50(4)**, 987-1007.

[4] Taylor S J 1986 Forecasting the volatility of currency exchange rates, *International Journal of Forecasting* **3(1)**, 159-170.

[5] French K R 1987 Expected stock returns and volatility, *Journal of Financial Econometrics* **19(1)**, 3-29.

[6] Chou R Y 1988 Volatility persistence and stock valuations: some empirical evidence using GARCH, *Journal of Applied Econometrics* **3(4)**, 279-294.

[7] Pagan A R and Schwert G W 1990 Alternative models for conditional stock volatility, *Journal of Econometrics* **45(1-2)**, 267-290.

[8] Bera A K and Higgins M L 1993 A survey of ARCH models: properties, estimation and testing, *Journal of Economic Surveys* **7(4)**, 305-366.

[9] Gouriéroux C 1997 *ARCH Models and Financial Applications*. Springer-Verlag.

[10] Vilasuso J 2002 Forecasting exchange rate volatility, *Economics Letters* **76(1)**, 59-64.

[11] Conrad C, Jiang F and Karanasos M 2003 Modelling and predicting exchange rate volatility via power ARCH models: the role of long-memory, *Working Paper*, University of Mannheim.

[12] Poon S H and Granger C W J 2003, Forecasting volatility in financial markets: a review, *Journal of Economic Literature* **41(2)**, 478-488.

[13] Dionísio A, Menezes R and Mendes D A 2007 On the integrated behaviour of non-stationary volatility in stock markets, *Physica A* **382(1)**, 58-65.

[14] Martens M and van Dick D 2007 Measuring volatility with realized range, *Journal of Econometrics* **138(1)**, 181-207.

[15] Kang S H, Kang S-M and Yoon S-M 2009 Forecasting volatility of crude oil markets, *Energy Economics* **31(1)**, 119-125.

[16] Vrught E B 2009 US and Japanese macroeconomic news and stock market volatility in Asia-Pacific, *Pacific-Basin Finance Journal* **17**, 611-627.

[17] Rodríguez-Villar M J 2010 *Volatility Models with Leverage Effect*, Ph.D. Thesis, Department of Statistics, Universidad Carlos III de Madrid.

[18] Vandewalle N and Ausloos M 1997 Coherent and random sequences in financial fluctuations, *Physica A* **246(3-4)**, 454-459.

[19] Nelson D B 1991 Conditional heteroskedasticity in asset returns: a new approach, *Econometrica* **59(2)**, 347-370.

[20] Hamao Y, Masulis N R and Ng V 1990 Correlations in price changes and volatility across international stock markets, *The Review of Financial Studies* **3(2)**, 281-307.

[21] Wei J K C, Liu Y-J, Yang C-C and Chaung G-S 1995 Volatility and price change spillover effects across the developed and emerging markets, *Pacific-Basin Finance Journal* **3(1)**, 113-136.

[22] Ferreira N B, Menezes R and Mendes D A 2007 Asymmetric conditional volatility in international stock markets, *Physica A* **382(1)**, 73-80.

[23] Bentes S and Menezes R 2013 On the predictability of realized volatility using feasible GLS, *Journal of Asian Economics* **28**, 58-66.

[24] Jarque C M and Bera A K 1987 A test for normality of observations and regression residuals, *International Statistics Review* **55(2)**, 163-172.

[25] Stambaugh R F 1999 Predictive regressions, *Journal of Financial Economics* **54(3)**, 375-421.

[26] Diebold, F X and Mariano R S 1995 Comparing predictive accuracy, *Journal of Business and Economic Statistics* **13(3)**, 253-263.

[27] Clements M P and Harvey D I 2007 Forecast combination and encompassing, in Mills, T.C. e K. Patterson (Eds.) *Handbook of Econometrics: Applied Econometrics* **2**, Palgrave.

[28] Clements M P and Harvey D I 2011 Combining probability forecasts, *International Journal of Forecasting* **27(2)**, 208-223.

[29] Harvey D I, Leybourne S J and Newbold P 1998 Tests for forecast encompassing, *Journal of Business and Economic Statistics* **16(2)**, 254-259.

[30] Harvey D I and Newbold P (2000), Tests for multiple forecast encompassing, *Journal of Applied Econometrics* **15(5)**, 471–482.

SPMCS2014 IOP Publishing
Journal of Physics: Conference Series **604** (2015) 012015 doi:10.1088/1742-6596/604/1/012015

Disordered random matrices

M. P. Pato

Instítuto de Física, Universidade de São Paulo, Caixa Postal 66318, 05314-970 São Paulo, S.P., Brazil

E-mail: mpato@if.usp.br

Abstract. A revision is made of the effect, in the statistical properties of the matrices of random matrix theory, of dividing them by a positive random variable. As the method preserves unitary invariance analitical derivations are easily performed. It is shown that the same procedure applied to the adjacency matrices of random graphs gives rise to a family of graphs that interpolates between the Erdös-Renyi and the scale-free model. Previous applications and new ones are discussed.

1. Introduction

After being linked to the manifestations of chaos in quantum physical systems[1], the classes of random matrices introduced by Wigner in the 50s have found a great success. This success generated a great activity and extensions and generalizations of that ensemble have occurred. To the description of the spectral statistical fluctuations observed in compound nuclei, Wigner adapted the Wishart ensemble well known to statisticians. Some of the extensions of the Gaussian ensemble also can be considered as applications of common processes in statistics, for instance, models to describe symmetry breaking have been constructed by adding two random matrices, one block diagonal and, the other, its complement[2].

The concept of disorder has been introduced in random matrix theories[3] to denote ensembles constructed by superimposing an external source of randomness to the internal fluctuations of their matrices. A random process in which there is a competition between two types of random variables is typical of disordered systems or, in the case of Ising models, spin glasses[4]. As the two types of randomness are independent, one can be kept quenched while the fluctuations of the other continue to operate. Several families of ensembles fit into this category, for instance, for the Wigner Gaussian case see Ref. [5, 6]; for the case of Wishart matrices see [7]; and for the case of random graphs see [3]. Disordered ensembles can be considered as a special case of the so-called superstatistics[8].

Here, a revision is made of the procedure of introducing disorder in random matrices focussing in new results that go beyond the previous applications made in modelling regimes in growth process[9] and in quantum chaos[10].

2. Disordered Gaussian matrices

Generating a new random variable by taking the ratio of two other random ones is a common procedure known by statisticians and, sometimes called randomization[11]. By taking a random matrix H_G of the Gaussian ensemble and dividing it by a positive random variable ξ, this procedure has been extended in [3] to the case of random matrices. A new ensemble of matrices given by

$$H = \frac{H_G}{\sqrt{\xi/\bar{\xi}}} \tag{1}$$

is constructed in which H_G are matrices whose elements have joint density probability distribution[12]

$$P_G(H) = \left(\frac{\beta}{2\pi}\right)^{f/2} \exp\left(-\frac{\beta}{2}\mathrm{tr}H^2\right), \tag{2}$$

while ξ is a positive random variable with density distribution $w(\xi)$ and first moment $\bar{\xi}$. In (2), f is the number $f = N + \beta N(N-1)/2$ of independent elements of Hermitian matrices of dimension N and β is the Dyson index $\beta = 1, 2, 4$ for GOE (real elements), GUE (complex elements) and GSE (quaternion elements). Here and in what follows the subindex G indicates Gaussian. In this way, an external source of randomness is superimposed to the fluctuations of the Gaussian matrix H_G. Here the disorder is represented by ξ which is the quenched variable in opposition to the randomness of the Gaussian matrices. In physical terms, the randomization procedure expressed by Eq. (1) leads to a disordered ensemble.

It is straightforward to deduce from Eqs. (1) and (2) that the disordered ensemble has density distribution

$$P(H) = \int_0^\infty d\xi w(\xi) \left(\frac{\beta\xi}{2\pi\bar{\xi}}\right)^{f/2} \exp\left(-\frac{\beta\xi}{2\bar{\xi}}\mathrm{tr}H^2\right), \tag{3}$$

an expression, Eq. (3), with characteristics of the so-called superstatistics[8], but, the fact that the two distributions appear normalized inside the integral, departs it from that scheme.

Turning now to eigenvalues and eigenvectors, we observe that Eq. (1) implies in H having the same eigenvectors of H_G. Therefore Eq. (1) can be diagonalized to give

$$D = \frac{D_G}{\sqrt{\xi/\bar{\xi}}}, \tag{4}$$

where the matrices D and D_G are diagonal matrices containing the respective eigenvalues, as a consequence, the relation

$$P\left(E_1, ...E_N\right) = \int_0^\infty d\xi w\left(\xi\right) \left(\alpha\xi/\bar{\xi}\right)^{\frac{N}{2}} P_G\left(x_1, ...x_N\right) \tag{5}$$

follows, where $x_i = \sqrt{\xi/\bar{\xi}}E_i$ and

$$P_G(x_1, ...x_N) = K_N^{-1} \exp\left(-\frac{\beta}{2}\sum_{k=1}^N x_k^2\right) \prod_{j>i} |x_j - x_i|^\beta \tag{6}$$

is the known normalized eigenvalue distribution of the Gaussian ensemble.

Eq. (5) means that measures of the generalized family can be calculated by weighting the corresponding measures of the Gaussian ensembles with the $w(\xi)$ distribution. For instance the eigenvalue density is expressed in terms of the Wigner's semi-circle law[12] as

$$\rho\left(E\right) = \frac{1}{\pi}\int_0^{\xi_{max}} d\xi w(\xi)(\xi/\bar{\xi})^{\frac{1}{2}}\sqrt{2N - \xi E^2/\bar{\xi}}. \tag{7}$$

where $\xi_{max} = 2\bar{\xi}N/E^2$. By making the substitution $v = \xi/\xi_{max}$ the density becomes

$$\rho(E) = \frac{(2N)^2\bar{\xi}}{\pi|E|^3}\int_0^1 d\xi w\left(\frac{2N\bar{\xi}v}{E^2}\right)\sqrt{v(1-v)}, \tag{8}$$

which, depending on the behavior of the function $w(\xi)$ at the origin, implies in a power-law decay. Ensembles showing heavy-tailed densities occurs, for instance, when $w(\xi)$ is the gamma distribution

$$w(\xi) = \exp(-\xi)\xi^{\bar{\xi}-1}/\Gamma(\bar{\xi}). \tag{9}$$

With variance $\sigma_\xi = \sqrt{\bar{\xi}}$ this distribution becomes more localized when $\bar{\xi}$ increases and, as a consequence, it should be expected that the Gaussian ensembles is recovered for large values of $\bar{\xi}$.

Another choice that has been investigated is the hyperbolic function

$$w(a,b,\lambda,\xi) = \frac{(b/a)^{\lambda/2}\xi^{\lambda-1}}{2K_\lambda(\sqrt{ab})}\exp\left[-\frac{1}{2}\left(\frac{a}{\xi}+b\xi\right)\right], \tag{10}$$

which, in contrast to Eq. (9), has all moments, and should produce a more regular behavior. The introduction in the Gaussian matrices of this disorder lead to a model that gives a good description of the intermediate statistics[13] exhibitted by some quantum physical systems[10].

3. A random walk

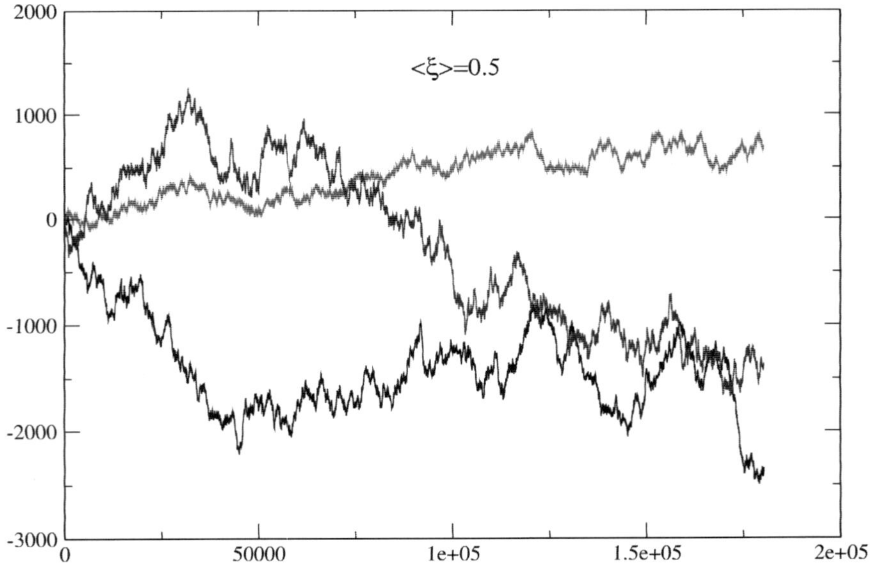

Figure 1. Random walk with correlated gaussian steps with $\xi = 0.5$.

Substituting Eq. (9) in Eq. (3), the integral in the external variable ξ can be performed to give

$$P(H,\bar{\xi}) = \left(\frac{\beta}{2\pi\bar{\xi}}\right)^{\frac{f}{2}}\frac{\Gamma\left(\bar{\xi}+\frac{f}{2}\right)}{\Gamma\left(\bar{\xi}\right)\left(1+\frac{\beta}{2\bar{\xi}}\mathrm{tr}H^2\right)^{\bar{\xi}+\frac{f}{2}}} \tag{11}$$

for the joint density distribution of matrix elements. Defining

$$\frac{1}{q-1} = \bar{\xi} + \frac{f}{2}, \text{ with } q > 1. \tag{12}$$

Eq. (11) is just Eq. (4) of [5, 6] which was derived using a generalized maximum entropy principle[14] with q being identified with Tsallis entropic parameter.

Keeping n matrix elements fixed and integrating the others, it is found that the probability of a set of $n \leq f$ matrix elements is given by an expression with the same structure of Eq. (11) that is

$$p(h_1, h_2, ..., h_n) = \left(\frac{\beta}{2\pi\bar{\xi}}\right)^{n/2} \frac{\Gamma\left(\bar{\xi} + \frac{n}{2}\right)}{\Gamma\left(\bar{\xi}\right)\left(1 + \frac{\beta}{2\bar{\xi}}\sum_i^n h_i^2\right)^{\bar{\xi}+\frac{n}{2}}} \tag{13}$$

where $h_i = H_{ij}$ for the diagonal and $h_i = \sqrt{2}H_{ij}$ for the off-diagonal elements. Eq. (13) makes evident that the matrix elements are correlated.

Eq. (1) makes straightforward to do numerical simulations in terms of the Gaussian matrices. However, matrices of the ensemble also can be generated by taking into account the correlations among their elements. To do this, consider the identity

$$p(h_1, ..., h_f) = p(h_1) \prod_{n=2}^{f} \frac{p(h_1, ...h_n)}{p(h_1, ...h_{n-1})} \tag{14}$$

where each fraction gives the conditional probability $P(h_n|h_1, ..., h_{n-1})$ for the nth element once the $n-1$ previous ones are given. Since

$$P(h_n|h_1, ..., h_{n-1}) = \sqrt{\frac{\beta}{2B_{n-1}\pi\bar{\xi}}} \frac{\Gamma(\bar{\xi} + \frac{n}{2})}{\Gamma(\bar{\xi} + \frac{n-1}{2})} \frac{1}{\left(1 + \frac{\beta h_n^2}{2B_{n-1}\bar{\xi}}\right)^{\bar{\xi}+n/2}}, \tag{15}$$

where

$$B_{n-1} = 1 + \frac{\beta}{2\bar{\xi}} \sum_{i=1}^{n-1} h_i^2, \tag{16}$$

the variables h_i sequentially can be sorted using, at each step, an univariate distribution. The sum of the elements thus generated, defines the random walk

$$S_n(\bar{\xi}) = \sum_{i=1}^{n} h_i. \tag{17}$$

Although, the sequence $h_1, h_2, ..., h_n$ is made of correlated variables, by Eq. (1), every matrix of the disordered ensemble should belong to a Gaussian distribution defined by a particular value of the external variable ξ. This means that the random walk is made of a set of correlated steps taken from a Gaussian distribution.

In Fig. 1 and Fig. 2, plots of calculated $S_n(\bar{\xi})$ shows the behavior of the random walk for the values $\bar{\xi} = 0.5$ and 100 respectively.

4. Fluctuations of the largest eigenvalue

The probability $E_{G,\beta}$ that the infinite interval (t, ∞) has no eigenvalues, that is, that the largest eigenvalue x_{max} be less than t has been calculated for the three classes of the Gaussian ensemble

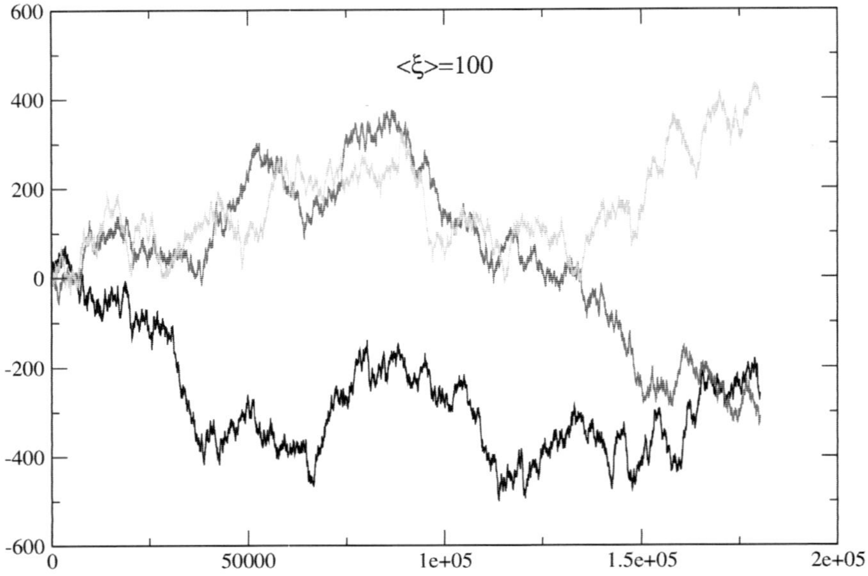

Figure 2. Random walk with Gausian correlated steps for $\xi = 100$.

[15] and found many applications[16]. At the edge of the spectrum, it is necessary to introduce the scaling

$$x = (2\sqrt{N} + \frac{s}{N^{1/6}})\sigma \tag{18}$$

where σ is the variance $\sigma = 1/\sqrt{2\beta}$ of the off-diagonal matrix elements. In this new scaled variable, s, the probability $E_{G,2}(s)$, for the unitary case, $\beta = 2$, is

$$E_{G,2}(s) = \exp\left[-\int_s^\infty (x-s)q^2(x)dx\right], \tag{19}$$

where $q(s)$ satisfies the Painlevé II equation

$$q'' = sq + 2q^3 \tag{20}$$

with boundary condition

$$q(s) \sim \mathrm{Ai}(s) \text{ when } s \to \infty, \tag{21}$$

where $\mathrm{Ai}(s)$ is the Airy function. For GOE ($\beta = 1$) and GSE ($\beta = 4$) are

$$[E_{G,1}(s)]^2 = E_{G,2}(s)\exp[-\mu(s)] \tag{22}$$

and

$$[E_{G,4}(s)]^2 = E_{G,2}(s)\cosh^2\frac{\mu(s)}{2} \tag{23}$$

where

$$\mu(s) = \int_s^\infty q(x)dx. \tag{24}$$

The above equations give a complete description of the fluctuations of the eigenvalues at the edge of the spectra of the Gaussian ensembles.

Turning now to the disordered case, the probability that the largest eigenvalue E_{max} is smaller than a given value t is obtained by averaging the above expression with the weight distribution $w(\xi)$ as

$$E_{\beta}\left(E_{max} < t\right) = \int_{0}^{\infty} d\xi w(\xi) E_{G,\beta}\left[S(\xi,t)\right], \tag{25}$$

where the argument of $S(\xi,t)$ is obtained by plugging in Eq. (18) the variance $\sigma(\xi) = \sqrt{\bar{\xi}/2\beta\xi}$ such that

$$S(\xi,t) = N^{1/6}\left[t\sqrt{\frac{2\beta\xi}{\bar{\xi}}} - 2\sqrt{N}\right]. \tag{26}$$

One successful application of the Tracy-Widom distributions was to growth processes[17]. Considering the evolution of a surface generated by an automata cellular model, analitically was found that the maximum point at the frontier follows the Tracy-Widom distribution. It was then also found that, in the presence of a random media, a transition occurs from the Tracy-Widom to a Gaussian regime[18]. Our disordered ensemble seems well fitted to model this transition. In fact, by introducing a scaling of the average $\bar{\xi}$ of Eq. (9) with the matrix size N, it was shown, in Ref. [9], that the largest eigenvalue, in the limit $N \to \infty$, undergoes the transition from Tracy-Widom to Gaussian. Moreover, a critical regime described by the convolution of the two limiting distributions has also been observed.

The fluctuations described by the Tracy-Widom distributions and its disordered, discussed above, conterpart, are well localized around the edge of the semi-circle density. Another kind of fluctuations would be very large and, in principle, very rare events in which deviations are of the order of the size of the spectrum. This kind of fluctuations have been calculated[19] including the probability that all eigenvalues be negative (or positive). In order to address the effect of disorder in these large fluctuations, we can put in Eq. (25) the distribution derived in Ref. [19] instead of the Tracy-Widom ones. Letting this approach to a future work, here we consider what happens in the limit $N \to \infty$ when the scaling of $\bar{\xi}$ is removed.

In this limit, compared with the function $w(\xi)$, and keeping the ratio $t/\sqrt{2N}$ fixed, the function $E_{G,\beta}$ approaches a step function centered at $\xi = 2N\bar{\xi}/t^2$. As a consequence, the probability distribution for the largest eigenvalue converges to

$$E_{\beta}\left(E_{max} < t\right) = \int_{2N\bar{\xi}/t^2}^{\infty} d\xi w\left(\xi\right) \tag{27}$$

with density distribution

$$\frac{dE_{\beta}\left(t\right)}{dt} = \frac{4N\bar{\xi}w(2N\bar{\xi}/t^2)}{t^3}. \tag{28}$$

Using Eq. (9), Eq. (28) becomes

$$\frac{dE_{\beta}\left(t\right)}{dt} = \frac{2\bar{\xi}^{\bar{\xi}}\exp(-\bar{\xi}/t^2)}{\Gamma(\bar{\xi})t^{2\bar{\xi}+1}}. \tag{29}$$

Notice that in (28) the variable t is the eigenvalue itself without any edge scaling. This means that, in this case, fluctuations become of the order of the size of the average ensemble spectrum. Eq. (29) resembles a Fréchet distribution, that is the distribution that describes extreme values at the edge of uncorrelated sequences with a power-law decay. However, the fact

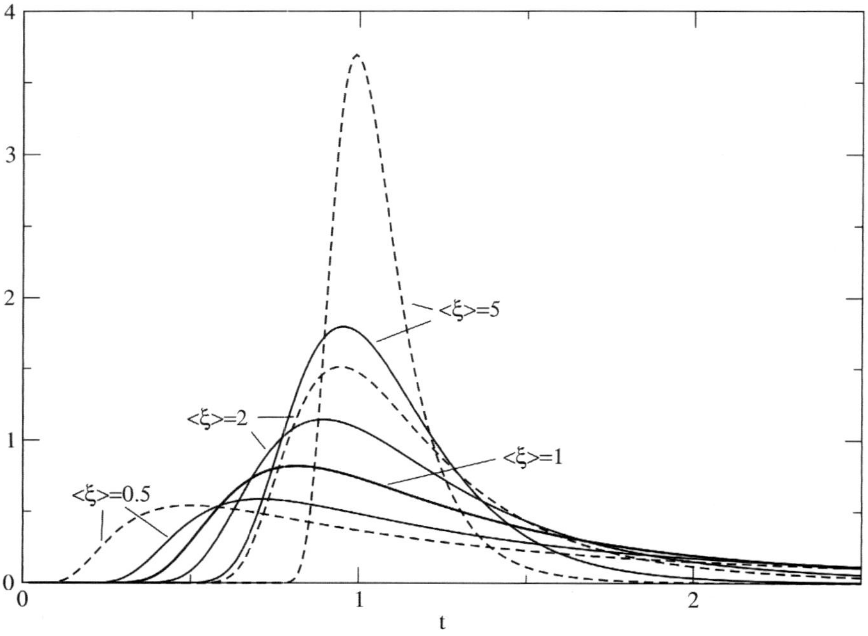

Figure 3. For indicated values of $\bar{\xi}$ the distributions of the largest eigenvalue of the disordered ensemble are compared with the Fréchet distribution with the same decay.

that the power of t in the exponent is fixed at the value two makes it a different distribution except for $\bar{\xi} = 1$ when it coincides with a Fréchet distribution. Therefore, Eq. (29) defines a long-tailed distribution of extreme values of a correlated set of points. The comparison with Fréchet distribution is shown in Fig. 3.

Another important point is that the distribution Eq. (29), and supposedly also the general one, Eq. (28), vanishes at the origin that is for $t = 0$. Therefore, the above approach cannot reproduce the results of Ref. [19]. In particular, it is remarkable that the introduction of disorder does not affect the probability of having all eigenvalues negative (or positive). In fact, putting $t = 0$ in Eq. (25), one concludes from Eq. (26) that the disordered distribution of the largest eigenvalue coincides with the Tracy-Widom distribution.

5. Bernoulli disordered random variable

One convenient way to characterize a Bernoulli variable of probability p is by the Logit function[20]

$$\alpha = \mathrm{Logit}(p) = \log\left(\frac{p}{1-p}\right), \tag{30}$$

where, as p is defined in the interval $[0,1]$, the domain of variation of α is $]-\infty,\infty[$. In terms of α, the probability is expressed as

$$p = \mathrm{Logit}^{-1}(\alpha) = \frac{1}{1+\exp(-\alpha)}. \tag{31}$$

Considering a set $a = (a_1, a_2, ..., a_f)$ of f independent Bernoulli variables with values $1(0)$, their joint density distribution is expressed in terms of the Logit function α as

$$P(a_1, a_2, ..., a_f) = \frac{1}{[1+\exp(\alpha)]^f} \exp\left(\alpha \sum_{i=1}^{f} a_i\right). \tag{32}$$

In particular, the probability that k elements of the set have value one while the others vanish is

$$\text{Prob}\left(\sum_{i=1}^{f} a_i = k\right) = \frac{f!}{k!(f-k)!} \frac{\exp(\alpha k)}{[1+\exp(\alpha)]^f} \tag{33}$$

If the Logit function is multiplied by the ratio $\xi/\bar{\xi}$ between a positive random variable ξ of density distribution $w(\xi)$ and its average, then the set a becomes a correlated set of disordered Bernoulli variables with joint density distribution

$$P(a_1, a_2, ..., a_f) = \int_0^\infty \frac{d\xi w(\xi)}{[1+\exp(\xi\alpha/\bar{\xi})]^f} \exp\left(\frac{\alpha\xi}{\bar{\xi}} \sum_{i=1}^{f} a_i\right). \tag{34}$$

The probability that k elements of the set have value one while the others vanish is, in this case, given by

$$\text{Prob}\left(\sum_{i=1}^{f} a_i = k\right) = \frac{f!}{k!(f-k)!} \int_0^\infty \frac{d\xi w(\xi)}{[1+\exp(\xi\alpha/\bar{\xi})]^f} \exp\left(\frac{\alpha\xi}{\bar{\xi}} k\right). \tag{35}$$

A graph is an array of N points (nodes) connected by edges, it can be represented by its adjacency matrix A whose elements A_{ij} have value 1(0) if the pair (ij) of nodes is connected (disconnected). The diagonal elements are taken equal to zero, i.e. $A_{ii} = 0$. A classical graph model was proposed by Erdös-Renyi (ER) in which there is a fixed probability p that a given pair of nodes is connected, independently of the others[22]. In this case, the adjacency matrix becomes a random matrix whose joint density distribution of elements is given as in Eq. (32) with $f = \frac{N(N-1)}{2}$. A disordered Erdös-Renyi then follows in which elements of the adjacency matrix has density distribution as in Eq. (34).

For a not too rarefied graph, the eigenvalue density of the eigenvalues of the adjacency matrix A of the ER model can be obtained from the moments of the trace of the powers of the matrix and one finds that it obeys the Wigner semi-circle law[21]

$$\rho_{ER}(E, \alpha) = \begin{cases} \frac{1}{2\pi\sigma^2}\sqrt{4N\sigma^2 - E^2}, & \text{if } |E| < \sqrt{4N\sigma^2} \\ 0, & \text{if } |E| > \sqrt{4N\sigma^2} \end{cases} \tag{36}$$

where σ^2 is the variance of the matrix elements given by

$$\sigma^2 = p(1-p) = \frac{1}{4\cosh^2(\alpha/2)}. \tag{37}$$

The above argument fails if $p \sim 1/N$ in which case deviations from the semi-circle appear[23, 24].

Immediately, the eigenvalue density of the ER disordered random graph is obtained by multiplying α by the external random variable ξ and integrating it, that is

$$\rho(E;\alpha) = \frac{2}{\pi} \int_0^{\xi_m} d\xi w(\xi) \cosh(\frac{\alpha\xi}{2}) \sqrt{N - \cosh^2(\frac{\alpha\xi}{2}) E^2}, \tag{38}$$

where

$$\xi_m = \frac{2}{\alpha}\cosh^{-1}(\frac{\sqrt{N}}{E}). \tag{39}$$

To proceed, we choose the weight distribution $w(\xi)$ to be given by Eq. (9). In Ref. [3] it is shown that with this choice the generalized model has features of a scale-free graph. In particular, the eigenvalue density exhibits a crossover from the Wigner semi-circle law to a distribution highly peaked with exponential tails.

In the same spirit, the degree distribution is given by Eq. (35)

$$P_k(\alpha, \bar{\xi}) = \frac{(N-1)!}{k!(N-1-k)!\Gamma(\bar{\xi})}\int_0^\infty d\xi \exp\left(-\xi\right)\xi^{\lambda-1}\frac{\exp\left(\frac{k\alpha\xi}{\xi}\right)}{\left[1+\exp(\frac{\alpha\xi}{\xi})\right]^{N-1}}. \tag{40}$$

The above integral is hard to evaluate when the number of nodes is large, however, as one is usually interested in large graphs that is, large values of N, the integral in (40) becomes suitable to an asymptotic evaluation. In fact, if we define

$$F(\xi) = \xi - (\lambda-1)\ln\xi - k\ln p(\xi) - (N-1-k)\ln[1-p(\xi)] \tag{41}$$

where

$$p(\xi) = \frac{1}{1+\exp(\frac{-\alpha\xi}{\xi})} \tag{42}$$

we deduce that

$$P_k(\alpha, \bar{\xi}) = \frac{(N-1)!}{k!(N-1-k)!}w(\xi_s)\left[p(\xi_s)\right]^k\left[1-p(\xi_s)\right]^{N-1-k}\sqrt{\frac{2\pi}{F''(\xi_s)}} \tag{43}$$

where ξ_s is the solution of the transcendental equation $F'(\xi_s) = 0$. This asymptotic expression allows an accurate way to calculate the degree distribution of large graphs.

In Fig. 4, degree distributions using the asymptotic calculations are shown for the values $\bar{\xi} = 1, 5, 10$. The scaling $p = N^{-z}$ has been used. It is clear the power-law behavior for $\bar{\xi} = 1$ while, for the other cases, there is a cut-off of the large connectivities.

6. Conclusions

In summary, a revision is made of the introduction of an external source of randomness superimposed to the internal fluctuations of random matrices of an ensemble. The method introduces correlations among the matrix ensembles and , as a consequence, the ergodic property (equivalence of spectral and ensemble averages) has to be abandoned. Therefore the new ensembles generated are more pertinent to cases in which only ensemble averages make sense as, for instance, the behavior of extreme eigenvalues. Another interesting application is to a random walk made of correlated Gaussian steps. Also in the case of random graph, the correlations introduced has an effect similar to that of the preferential attachement of the scale-free graphs[21].

Acknowlegments

This work is supported by the Brazilian agencies CNPq and FAPESP.

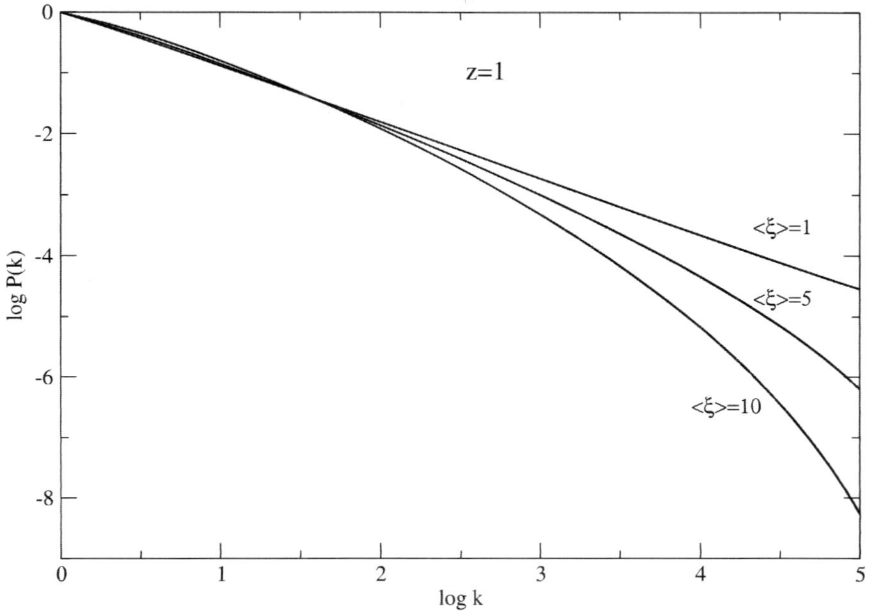

Figure 4. Asymptotic degree distributions for $\bar{\xi} = 1, 5, 10$ assuming thescaling $p = N^{-z}$.

References

[1] Bohigas O., Giannoni M J and Schmit C 1984 *Phys. Rev. Lett.* **52** 1; Sieber M and Richter K 2001 *Physica Scripta* T **90** 128; Müller S, Heusler S, Braun P, Haake F and Altland A 2005 *Phys. Rev.* E **72** 046207; Heusler S, Müller S, Altland A, Braun P and Haake F 2007 *Phys. Rev. Lett.* **98** 044103

[2] Hussein M S and Pato M P 1993 *Phys. Rev. Lett.* **70** 1089

[3] Bohigas O, de Carvalho J X and Pato M P 2006 *Phys. Rev.* E **77** 011122

[4] Mézard M, Parisi G and Virasoro M 1987 *Spin glasses theory and beyond* (Singapore: World Scientific)

[5] Bertuola A C, Bohigas O and Pato M P 2004 *Phys. Rev.* E **70** 065102(R)

[6] Toscano F, Vallejos R O and Tsallis C 2004 *Phys. Rev.* E **69** 066131; Abul-Magd A Y 2005 *Phys. Rev.* E **71** 066207; Muttalib K A and Klauder J R 2005 *Phys. Rev.* E **71** 055101(R)

[7] Akemann G and Vivo P 2008 *J. Stat. Mech.* P09002; Abul-Magd A Y, Akemann G and Vivo P 2009 *J. Phys.* A: Math. Theor. **42** 175207; Akemann G, Fischmann J and Vivo P 2010 *Physica A* **389** 2566

[8] Beck C and Cohen E G D 2003 *Physica A* **322** 267

[9] Bohigas O, de Carvalho J X and Pato M P 2009 *Phys. Rev.* E **79** 031117

[10] Bohigas O and Pato M P 2011 *Phys. Rev.* E **84** 031121

[11] Feller W 1970 *An Introduction to Probability Theory and Its Applications* (New York: John Wiley and Sons, Inc.)

[12] Mehta M L 2004 *Random Matrices* (Elsevier Academic Press, 3nd Ed.)

[13] Bohigas O and Pato M P 2006 *Phys. Rev.* E **74** 036212

[14] Tsallis C, Mendes R S and Plastino A R 1998 *Physica A* **261** 534

[15] Tracy C A and Widom H *Commun. Math. Phys.* **159** 151 (1994) and **177** 727 (1996)

[16] Tracy C A and Widom H 2002 *Proceedings of the ICM*, Beijing **1** 587-96

[17] Gravner J, Tracy C A and Widom H 2001 *J. Stat. Phys.* **102** 1085

[18] Gravner J, Tracy C A and Widom H 2002 *Ann. of Prob.* **30** 1340; 2002 *Commun. Math. Phys.* **229** 433 Johansson K 2007 *Prob. Theo. and Rel. Fields* **138** 75

[19] Dean D S and Majumdar S N 2006 *Phys. Rev. Lett.* **97** 160201

[20] Cramer J S 2003 *The origins and development of the logit model* (Cambridge: Cambridge UP)

[21] Albert R and Barabási A-L/ 2002 *Rev. Mod. Phys.* **74** 47

[22] Erdös P and Renyi A 1959 *Publ. Math. Debrecen* **6** 290

[23] Farkas I J, Derényi I, Barábasi A-L and Vicsek T 2001 *Phys. Rev.* E **64** 026704

[24] Semerjian G and Cugliandolo L F 2002 *J. Phys. A: Math. Gen.* **35** 4837

The Society of Brains: How Alan Turing and Marvin Minsky Were Both Right

Zbigniew R. STRUZIK[1,2,3]

[1] RIKEN Brain Science Institute, 2-1 Hirosawa, Wako-shi 351-0198, Japan
[2] Graduate School of Education, The University of Tokyo, 7-3-1 Hongo, Bunkyo-ku, Tokyo 113-0033, Japan
[3] Institute of Theoretical Physics and Astrophysics, The University of Gdańsk, Wita Stwosza 57, PL-80-952 Gdansk, Poland

E-mail: z.r.struzik@p.u-tokyo.ac.jp

Abstract. In his well-known prediction, Alan Turing stated that computer intelligence would surpass human intelligence by the year 2000. Although the Turing Test, as it became known, was devised to be played by one human against one computer, this is not a fair setup. Every human is a part of a social network, and a fairer comparison would be a contest between one human at the console and a network of computers behind the console.

Around the year 2000, the number of web pages on the WWW overtook the number of neurons in the human brain. But these websites would be of little use without the ability to search for knowledge. By the year 2000 Google Inc. had become the search engine of choice, and the WWW became an intelligent entity. This was not without good reason. The basis for the search engine was the analysis of the 'network of knowledge'. The PageRank algorithm, linking information on the web according to the hierarchy of 'link popularity', continues to provide the basis for all of Google's web search tools. While PageRank was developed by Larry Page and Sergey Brin in 1996 as part of a research project about a new kind of search engine, PageRank is in its essence the key to representing and using static knowledge in an emergent intelligent system.

Here I argue that Alan Turing was right, as hybrid human-computer internet machines have already surpassed our individual intelligence - this was done around the year 2000 by the Internet - the socially-minded, human-computer hybrid Homo computabilis-socialis. Ironically, the Internet's intelligence also emerged to a large extent from 'exploiting' humans – the key to the emergence of machine intelligence has been discussed by Marvin Minsky in his work on the foundations of intelligence through interacting agents' knowledge.

As a consequence, a decade and a half decade into the 21st century, we appear to be much better equipped to tackle the problem of the social origins of humanity – in particular thanks to the power of the intelligent partner-in-the-quest machine, however, we should not wait too long...

1. Introduction

Our ability to form a functional unity with the tools we use is almost proverbial. And indeed we are particularly good at making and using machines. This ability to form functional 'imaginary' unity scales both up and down. It scales down to brain modules and neuronal assemblies, possibly to individual cell interactions. And it scales up to interpersonal relations, social assemblies of individuals and organisational modules and structures – along the dimension of social network interactions. Analysing the properties and understanding the mechanisms of emergence of such functional networks (of networks) of interactions may be essential in grasping the origins and basis of human conscious intelligence.

2. Networks of Brain Networks

The neurons and glia cells in our brains form a hybrid network of networks and an apparent, albeit as of today still only speculative, scaling continuum of social interactions with the outside world. Throughout the evolutionary refinement leading to the modern organisation of our (Homo Sapiens) complex society, one mechanism appears dominant. The organisation of our brains adapts to the increasing complexity of societal organisation. As this complexity of societal organisation increases, apparently so does that of our self-aware, conscious existence.

The amazingly complex working of hybrid brain networks is far from being understood. Yet, the bonding concept behind our interpretation of the guiding principle behind the adaptive complexity of this system is its ability 'intelligently' to process information. Even the neuronal networks grown in vitro are capable of developing 'intelligent' communication resembling that of human social interactions. Orlandi et al. [1] recently studied the mechanism of burst propagation in cultured neuronal networks observed with high-resolution calcium imaging and in silico. They identied what could be described as an emerging 'functional adaptive network' - a set of points specific to each culture and selected by a non-trivial interplay between the dynamics and the topology of the network. On the basis of the statistics of avalanche size at different scales, they have shown that one may identify different effective networks which decompose the dynamics into separate layers. The focal points which appeared to be most influential in the global dynamics did not exactly follow local properties of the original or the effective network for large avalanches, but resulted from complex patterns of propagation. This mechanism appears to have direct correspondence with that of rumour propagation in social networks, where the role of the integrate-and-fire response is played by the so-called illusion-of-truth effect, that is, the requisite of repeated inputs to grant credibility, before propagation. Accordingly, not only the rumour activity network will differ from the underlying social network, but the points of rumour ignition will in general depart from the actual community structure of both the social and the effective networks.

Such functional mapping may exist at the level of higher brain networks. Indeed, neuronal networks of our brains have an amazing capability to form a functional unity with the tools we create, including the 'instruments' of our societal, religious and cultural systems. One such mechanism, of 'mirror neurons' claimed to have been discovered in macaques, shows that premotor and parietal cortical areas are not only involved in executing ones own movement, but are also active when observing the action of others. To date there is, however, relatively weak evidence for the existence of a circuit with 'mirror properties in humans, such as that described in monkeys [2]. Although debates about the evolution of the mirror neuron system imply that it is an adaptation for understanding of actions, an alternative, simpler explanation suggests that mirror neurons may be a by-product of associative learning. Heyes [3] argues that the mirror neuron system is a product, as well as a process, of social interaction. The associative account implies that mirror neurons come from sensorimotor experience, and that much of this experience is obtained through interaction with others.

While the 'mirror neuron network' primarily aims at explaining sensorimotor behaviour, the paradigm and research questions of 'neuroeconomics' address the greater concept of social mechanism and choices. However, they may lead to a paradox, as other primates are likely better than us at survival games [5]. What they apparently miss are the multi-dimensions and multi-scales of both the social and temporal horizons and the associated complexity of conscious strategy making. Humans are particularly good at deception games, rumour spreading and social bonding through 'gossip'. These and other social tools have been highly developed by

Homo sapiens and constitute the intrinsic fabric of our social intelligence. Furthermore, humans have perfected 'scaling' principles in applying these social interaction practices, through which they apply the rules to social structures of arbitrary size. One recent study of interest in this context analysed the bonding effect of pro-social lies as opossed to destructive anti-social lies [4]. Indeed, such manipulative practices are considered to be distinctly 'human'. Yet, the possibility of simulating these on 'non-human' models and von Neumann computer architectures suggests something quite to the contrary - that our treasured 'intelligent' behaviours may be shared by any social group of machines.

There are recent ongoing efforts to model and study through simulation in the context of the increasingly popular discipline of 'socio-physics'. Indeed, the agents which are studied do posess only minimal 'intelligence' yet their collective behaviour may exhibit capacities to form emergent intelligent behaviours – which may also be scaling.

Intrapolating (extrapolating downwards the scales) social bonding mechanisms is at this point highly speculative. Yet, one cannot exclude the possibility that the neuronal networks could behave just like ourselves and 'play' 'mirroring' - or 'aping' and 'immitation', 'deception', pro- and anti-social 'rumour' and 'gossip spreading' games. While the similarity of the behaviour observed by Orlandi et al. [1] in neuronal networks and our social interactions may be strictly formal, the possibility that such behaviour scales across the 'brain-society-barrier' is captivating and deserves future formal and experimental exploration.

3. Homo computabilis-socialis

The title of this article purposefully resembles that of the Marvin Minsky's highly acclaimed book. Indeed, as also discussed above, not neccesarily intelligent agents may develop 'machine' intelligence through network-scale interactions – which is also the central concept behind Minsky's work[1]. However, there is an irresistibly tantalising twist to this idea, which is also somewhat petrifying. The concept of mindless agents can be – and indeed is – explored in the context of socio-physics to model ourselves i.e. Homo sapiens involved in a network of interactions leading to emergent intelligence.

For this reason, the apparent re-use of the title is to echo the famous work and the concept but it is not supposed to mimic the subject of the book - at least not within this section. Here, the title of the book is evoked in a new meaning, as here humans play the role of not necessarily intelligent 'agents' involved in social interactions and advanced into an intelligent society. This evolution process is reciprocal in that the agents locally 'mirror' to a certain degree the intelligence of the societal network in which they are embedded. Such 'mirroring' or social subgroup formation has been discussed in the previous section in the example context of bonding mechanisms of information spreading. The mechanisms involved are, however, multi-level and multi-modal and these have been extensively studied in the field of human indiviudual psychology and reciprocical social behaviours. There would be little to add to these well established insights if we were to evaluate the 'intelligence' of human beings. But we are not – in his well-known prediction, Alan Turing stated that computer intelligence would surpass human intelligence by the year 2000. Although the Turing Test, as it became known, was devised to be played by one human against one computer, this is not a fair setup. Every human is a part of a social network, and a more fair comparison is that between one human at the console and a network of computers behind the console.[2]

Towards the year 2000, the number of web pages on the WWW overtook the number of neurons in the human brain. But these websites would be of little use without the ability to

[1] Minsky postulated that human intelligence arises from interactions of mindless 'agents' as constituting a 'society of mind', hence the title.[6]

[2] The Turning Test was debatably passed by a stand-alone computer during the 2014 University of Reading competition. This test was, however, subject to constraints and has been criticised.

search for knowledge. By the year 2000 Google Inc. became the search engine of choice [7] and the WWW became an intelligent entity. This was not without good reason. The basis for the search engine was the analysis of the 'network of knowledge'. The PageRank algorithm, linking information on the web according to the hierarchy of 'link popularity', continues to provide the basis for all of Google's web search tools [8]. While PageRank was developed by Larry Page and Sergey Brin in 1996 as part of a research project about a new kind of search engine, the idea of formulating a link analysis problem as an eigenvalue problem was apparently first suggested in 1976 by Gabriel Pinski and Francis Narin, in their work in the context of the 'scientometrics' discipline of ranking scientific journals [9].

While the Pinsky and Narin were among the pioneers of 'scientometrics', the discipline is now flourishing in the modern physics of social interactions, in particular in more recent efforts aiming at modelling and characterising knowledge flow and innovation emergence. Yet, while physics provides models and experiments, these efforts have remained far behind an unprecedented development in the history of mankind and apparently of any kind of intelligent life of which we are awere. Indeed – Alan Turing was right, as hybrid human-computer internet machines powered by 'enslaved' humans feeding them with information and interacting with them have already surpassed our individual intelligence. This, coincidentally or not, has happened around the year 2000 with the Internet – the socially-minded, human-computer hybrid Homo computabilis-socialis.

Through engaging unparalleled conjoint effort of neuroscience, physics, psychological and social sciences, we are now much better equipped to tackle the problem of the social origins of humanity. Yet, as of today, we may still be unable to obtain sufficient insight to understand the very basis of our individual and social conscious intelligence. We cannot, however, delay much longer with making the understanding our own human nature a priority. Paradoxically, the machines we produce and with which we interact may understand us before we do.

References

[1] Orlandi J G, Soriano J, Alvarez-Lacalle E, Teller S and Casademunt J 2013 Noise focusing and the emergence of coherent activity in neuronal cultures *J. Nature Phys.* **9** 582-90

[2] Turella L, Pierno A C, Tubaldi F and Castiello U 2009 Mirror neurons in humans: Consisting or confounding evidence? *Brain & Language* **108** 10-21

[3] Heyes C 2010 Where do mirror neurons come from? *Neurosci. Biobehav. Rev.* **34** 575-83

[4] Iiguez G, Govezensky T, Dunbar R, Kaski K and Barrio R A 2014 Effects of Deception in Social Networks *Preprint* arXiv:1406.0673

[5] Martin C F, Bhui R, Bossaerts P, Matsuzawa T and Camerer C 2014 Chimpanzee choice rates in competitive games match equilibrium game theory predictions *Scientific Reports* **4** 5182

[6] Minsky M 1988 *The Society of Mind* (New York: Simon and Schuster)

[7] Rosenberg S 1998 Yes, there is a better search engine. While the portal sites fiddle, Google catches fire *Salon.com*

[8] Brin S and Page L 1998 The anatomy of a large-scale hypertextual Web search engine *Computer Networks and ISDN Systems* **30** 107-17

[9] Pinski B and Narin F 1976 Citation influence for journal aggregates of scientific publications: Theory, with applications to the literature of physics *Information Processing and Management* **12** 297-312

Entropy minimization within the maximum entropy approach

E.V. Vakarin, J.P. Badiali

IRCP UMR 8247 CNRS / ENSCP Chimie ParisTech, 11 rue P. et M. Curie, 75231 Cedex 05, Paris, France

E-mail: eduard.vakarin@upmc.fr

Abstract. A procedure of the minimization of the maximum entropy with respect to an external driving force μ is proposed. In application to coupled dynamic-stochastic systems such an approach allows one to reduce the uncertainty in estimating the dynamic counterpart (a probe or a model) parameters. In its turn this permits to estimate an optimal driving path giving a maximal information on the probability distribution $f(x|\mu)$ of the stochastic counterpart with a given probe/model $\theta(\mu|x)$. It is found that the functional form of the model should be similar to the observed/measured one $\theta(\mu)$, while the minimum uncertainty is reached when the distribution becomes independent of the driving μ.

1. Introduction

The term *Complex Systems* has emerged as a recognized field of research in many domains (from physics to biology). Nevertheless a strict definition of the complexity is still lacking. However there are several common generic features such as multicomponent nature, emergence, non-linear or even feedback relations among the components. In this situation one has to work under conditions of incomplete or partial information on the system. Thus a deterministic description becomes almost impossible and one has to resort to statistical/probabilistic methods. One of them is the maximum entropy approach[1] which is extensively invoked in the solution of inference problems appearing in various fields . The approach allows one to calculate a probability distribution of relevant variables from an incomplete knowledge of the microscopic mechanisms governing the system evolution. The central idea is to find a least biased distribution consistent with the information at hand. Therefore, the scheme operates with two main ingredients: the entropy measure (that estimates the uncertainty associated with the probability distribution) and the constraints encoding the available information. In this context various entropy functionals are introduced (see [2] for a recent review) in order to derive various non-exponential probability distributions observed in nature. These are generalizations of the famous Shannon form (4) that appears from the generalized forms as some limiting case. However, many of these generalized entropies are non-additive for statistically independent subsystems. This fact with its consequences is still a subject of extensive debates [2, 3].

In a situation when many generalized entropy forms are being proposed [4] it might seem that the maximum entropy approach suffers from an ambiguity [5]: "in a sense that any probability distribution seems to be derivable from the maximization of any entropic measure if an appropriate constraint is used". In this respect it has been argued [5] that such an ambiguity does not appear when one searches for a parametric family of distributions $f(x|\mu)$. Here x is a relevant fluctuating variable and μ is an external parameter independent of x. For instance, such a parameter could enter through an observed mean value $\theta(\mu)$ [6] or the distribution width varying in response to some external perturbation μ, like in the process of insertion into a host matrix [7]. On the other hand, we have to note that we are not free to modify the constraints without logical necessity because they contain the information (or our theoretical model) on the processes running in the system of interest.

Since we deal with a parametrized distribution $f(x|\mu)$ its entropy $H(\mu)$ is also a function of the parameter μ. This opens up a possibility to search for a minimum of $H(\mu)$ as a function of the driving μ. In other words, with a given observable $\theta(\mu)$ and its model estimation $\theta(\mu|x)$ we may try to find an "optimal" set of parameters reducing the uncertainty on the behavior of $f(x|\mu)$. This idea was briefly sketched in [8].

In this paper we explore such a possibility in application to coupled dynamic-stochastic systems [9, 10, 11], where the dynamic counterpart may serve as a probe for determining the behavior $f(x|\mu)$ of the stochastic subsystem. In this way the maximum entropy approach allows one to find a functional form of the distribution $f(x|\mu)$ while the subsequent minimization of the maximized entropy permits to reduce the uncertainty on the relevant parameters.

2. Maximum entropy approach

Let us first summarize some basic principles of the maximum entropy approach. In the light of what is discussed above, we consider a parametrized entropy functional $H(\mu)$ of a quite general form [8]

$$H(\mu) = D\left(\Phi\left[f(x|\mu)\right]\right) \tag{1}$$

Here H is simultaneously a function of the parameter μ and a functional of the parametrized distribution $f(x|\mu)$, $D(t)$ is some function with appropriate behavior. The functional $\Phi\left[f(x|\mu)\right]$ can be represented as

$$\Phi\left[f(x|\mu)\right] = \int dx \varphi\left(f(x|\mu)\right), \tag{2}$$

where $\varphi(p)$ is another function ensuring suitable overall properties (continuity, concavity,...). Combining eqs. (1) and (2) we arrive at

$$H(\mu) = D\left(\int dx \varphi\left(f(x|\mu)\right)\right) \tag{3}$$

It is clear that practically all generalized entropy functionals appearing in the current literature can be deduced from the definition above (except the forms involving the distribution derivatives like the Fisher form). For instance, for $D(t) = t$, $\varphi(p) = -p \ln p$ we recover the Shannon entropy

$$H_S(\mu) = -\int dx f(x|\mu) \ln f(x|\mu) \tag{4}$$

Let us consider a problem of maximizing $H(\mu)$ under two natural constraints - the normalization

$$\int dx f(x|\mu) = 1 \tag{5}$$

and some known average $\theta(\mu)$ of a conditional response function $\theta(\mu|x)$

$$\theta(\mu) = \overline{\theta(\mu|x)} = \int dx f(x|\mu)\theta(\mu|x). \tag{6}$$

Here the overbar denotes the corresponding average taken with the distribution $f(x|\mu)$. Usually the average $\theta(\mu)$ corresponds to the observed behavior (e.g. experimental data) and the conditional function $\theta(\mu|x)$ represents our theoretical estimation (a model) of this behavior for a given state x of the random environment. The latter is characterized by a probability distribution $f(x|\mu)$ depending on the "driving" parameter μ (for instance, this could be an external field, pressure or chemical potential).

Thus we have to find a variation of the following Lagrangian

$$L = H(\mu) + \kappa \left[\int dx f(x|\mu)\theta(\mu|x) - \theta(\mu) \right] + \lambda \left[\int dx f(x|\mu) - 1 \right], \tag{7}$$

where κ and λ are the Lagrange multipliers which should be found from the constraints (5) and (6). The variation procedure $\delta L/\delta f(x|\mu) = 0$ leads to

$$\frac{\partial D\left(\Phi\left[f(x|\mu)\right]\right)}{\partial \Phi\left[f(x|\mu)\right]} \frac{\partial \varphi\left(f(x|\mu)\right)}{\partial f(x|\mu)} = -\kappa\theta(\mu|x) - \lambda \tag{8}$$

In the case of the Shannon entropy $H(\mu) = H_S(\mu)$ (4) this procedure leads to an apparently exponential distribution

$$f(x|\mu) = \frac{\exp(-\kappa\theta(\mu|x))}{\int dx \exp(-\kappa\theta(\mu|x))} \tag{9}$$

Nevertheless, if the relevant variable is x, then the actual form of the distribution (9) depends on the form of the constrained function $\theta(\mu|x)$. For instance, as we have demonstrated [10, 11], if $\theta(\mu|x)$ is of logarithmic form, then (9) transforms into a Γ-distribution. Physically this corresponds to a non-equilibrium stationary state whose thermodynamic entropy is held at a given distance from the maximum that corresponds to the equilibrium. For $\theta(\mu|x) = g(\mu)x^\alpha$ we obtain from (9) a stretched exponential distribution. This means that we can associate exponential distributions to the Shannon entropy only in the case of linear constraint, i.e. $\theta(\mu|x) = g(\mu)x$.

3. Minimal maximized entropy

As is discussed above we are dealing with the parametrized entropy $H(\mu)$. Let us consider now the variation of $H(\mu)$ with respect to the parameter μ. From eqs. (1) and (2) we obtain

$$\frac{\partial H(\mu)}{\partial \mu} = \frac{\partial D\left(\Phi\left[f(x|\mu)\right]\right)}{\partial \Phi\left[f(x|\mu)\right]} \frac{\partial \Phi\left[f(x|\mu)\right]}{\partial \mu}; \qquad \frac{\partial \Phi\left[f(x|\mu)\right]}{\partial \mu} = \int dx \frac{\partial \varphi\left(f(x|\mu)\right)}{\partial f(x|\mu)} \frac{\partial f(x|\mu)}{\partial \mu} \tag{10}$$

Here $\partial D\left(\Phi\left[f(x|\mu)\right]\right)/\partial \Phi\left[f(x|\mu)\right]$ could be a function of μ. Anyway it does not depend on x since $\Phi\left[f(x|\mu)\right]$ is a functional of the probability distribution and simultaneously a function of μ. Therefore, combining these two equations we obtain

$$\frac{\partial H(\mu)}{\partial \mu} = \int dx \frac{\partial D\left(\Phi\left[f(x|\mu)\right]\right)}{\partial \Phi\left[f(x|\mu)\right]} \frac{\partial \varphi\left(f(x|\mu)\right)}{\partial f(x|\mu)} \frac{\partial f(x|\mu)}{\partial \mu} \tag{11}$$

Using now the condition (8) for the maximum entropy we arrive at

$$\frac{\partial H(\mu)}{\partial \mu} = -\kappa \int dx \theta(\mu|x) \frac{\partial f(x|\mu)}{\partial \mu} - \lambda \int dx \frac{\partial f(x|\mu)}{\partial \mu} \tag{12}$$

It is clear that the last term disappears because of the normalization condition (5) and we finally get the maximum entropy rate as

$$\frac{\partial H(\mu)}{\partial \mu} = -\kappa \int dx \theta(\mu|x) \frac{\partial f(x|\mu)}{\partial \mu}. \tag{13}$$

Now differentiating the constraint (6) we obtain

$$\frac{\partial\theta(\mu)}{\partial\mu} = \int dx \frac{\partial\theta(\mu|x)}{\partial\mu} f(x|\mu) + \int dx\theta(\mu|x)\frac{\partial f(x|\mu)}{\partial\mu} \tag{14}$$

Combining these two equations we arrive at

$$\frac{\partial H(\mu)}{\partial\mu} = -\kappa\left[\frac{\partial\theta(\mu)}{\partial\mu} - \overline{\frac{\partial\theta(\mu|x)}{\partial\mu}}\right] \tag{15}$$

Therefore, we have a "universal" relation among the entropy rate, the observed behavior $\theta(\mu)$ and the average of the model estimation $\theta(\mu|x)$. This relation does not depend on the form of the entropic functional (the form (1) is quite general) and thus could be helpful for drawing conclusions of general validity. A brief discussion of (15) for the case of the Shannon form can be found in [10, 11]. Anyway it is clear that the behavior of $H(\mu)$ depends on what is known/measured $\theta(\mu)$ and on our conditional estimation $\theta(\mu|x)$. Thus it seems logically acceptable to search for a way of reducing the ambiguity. Namely, one can analyze the conditions

$$\frac{\partial H(\mu)}{\partial\mu} = 0, \qquad \frac{\partial^2 H(\mu)}{\partial\mu^2} > 0 \tag{16}$$

for a minimum of $H(\mu)$. In such a way we can find some specific values of the parameters at which the uncertainty on the values of the relevant variable x is minimal. Such a procedure might help to determine which "driving path" is more informative. Let us illustrate this on a simple example. Suppose that there are two relevant choices:

$$\theta(\mu) = A\mu^q \tag{17}$$

$$\theta(\mu|x) = x\mu^\alpha \tag{18}$$

, where the value of α is to be estimated. Maximization of the Shannon entropy (4) under these conditions leads to

$$f(x|\mu) = \frac{\mu^{(\alpha-q)}}{A}\exp\left[-x\frac{\mu^{(\alpha-q)}}{A}\right] \tag{19}$$

Estimating the terms in the relation (15) we get

$$\frac{\partial\theta(\mu)}{\partial\mu} = Aq\mu^{q-1} \tag{20}$$

$$\overline{\frac{\partial\theta(\mu|x)}{\partial\mu}} = A\alpha\mu^{q-1} \tag{21}$$

Analyzing the first condition in (16) we arrive at $\alpha \to q$. From the second condition of (16) we conclude that $\alpha \to q$ from above or below, depending on the value of q. Based on these arguments and returning back to (19) one can easily see that the minimum uncertainty conditions point up towards choosing the model close to the observed behavior $\theta(\mu)$. In this way the probability distribution becomes independent of the driving μ.

Thus, the minima of $H(\mu)$ occurring at $\mu_1, ..., \mu_n$ correspond to the minima of the observed function $\theta(\mu)$ and this is equivalent to a minimization of the distribution width with fixing the positions of the distribution maxima at $x = \mu_1, ..., \mu_n$. Therefore, equation (15) together with the conditions (16) allow one to translate the specificities of the observed behavior $\theta(\mu)$ into the amount of information on the relevant random variable x. Obviously an interpretation of this correspondence depends on the nature of processes which are encoded in the function appearing in the constraint (6).

4. Conclusion

A minimization procedure of the maximum entropy of a quite general entropic functional with respect to an external driving force μ is proposed. It is demonstrated that for a conditional or parametrized probability distribution $f(x|\mu)$ there is a "universal" relation among the entropy rate and the functions appearing in the constraint. This relation allows one to translate the specificities of the observed behavior $\theta(\mu)$ into the amount of information on the relevant random variable x at different values of the parameter μ. It is clear that we are not free to change the constraint in *ad hoc* manner without some physical (or logical) reasons because it encodes our partial information $\theta(\mu)$ into what we are supposed to know $(\theta(\mu|x))$ and what we are searching for $(f(x|\mu))$ In application to coupled dynamic-stochastic systems such an approach allows one to reduce the uncertainty in estimating the dynamic counterpart (a probe or a model) parameters. In its turn this permits to estimate an optimal driving path giving a maximal information on the probability distribution $f(x|\mu)$ of the stochastic counterpart with a given probe/model $\theta(\mu|x)$. It is found that the functional form of the model should be similar to the observed/measured one $\theta(\mu)$, while the minimum uncertainty is reached when the distribution becomes independent of the driving μ.

References

[1] Jaynes E T 2006 *Probability Theory, The Logic of Science* (Cambridge: Cambridge University Press)
[2] Abe S and Okamoto Y 2001 *Nonextensive Statistical Mechanics and its Applications* (Berlin: Springer-Verlag)
[3] Beck C 2009 *Contemp. Phys.* **50** 495
[4] Abe S and Thurner S 2008 *Europhys. Lett.* **81** 10004
[5] Plastino A R, Plastino A and Soffer B H 2007 *Phys. Lett.* A **363** 48
[6] Vakarin E V, Duda Y and Badiali J P 2006 *J. Chem. Phys.* **124** 144515
[7] Garcia-Belmonte G, Vakarin E V, Bisquert J and Badiali J P 2010 *Electrochim. Acta* **55** 6123
[8] Vakarin E V and Badiali J P 2011 *Mod. Phys. Lett.* B **25** 1821
[9] Beck C 2001 *Phys. Rev. Lett.* **87** 180601
[10] Vakarin E V and Badiali J P 2006 *Phys. Rev.* E **74** 036120
[11] Vakarin E V and Badiali J P 2006 *AIP Conf. Proc.* **872** 411

Pascal (Yang Hui) triangles and power laws in the logistic map

Carlos Velarde[1], **Alberto Robledo**[2]

[1] Instituto de Investigaciones en Matemáticas Aplicadas y Sistemas, Universidad Nacional Autónoma de México, Mexico City, Mexico

[2] Instituto de Física y Centro de Ciencias de la Complejidad, Universidad Nacional Autónoma de México, Mexico City, Mexico

E-mail: robledo@fisica.unam.mx

Abstract. We point out the joint occurrence of Pascal triangle patterns and power-law scaling in the standard logistic map, or more generally, in unimodal maps. It is known that these features are present in its two types of bifurcation cascades: period and chaotic-band doubling of attractors. Approximate Pascal triangles are exhibited by the sets of lengths of supercycle diameters and by the sets of widths of opening bands. Additionally, power-law scaling manifests along periodic attractor supercycle positions and chaotic band splitting points. Consequently, the attractor at the mutual accumulation point of the doubling cascades, the onset of chaos, displays both Gaussian and power-law distributions. Their combined existence implies both ordinary and exceptional statistical-mechanical descriptions of dynamical properties.

1. Introduction

The logistic map has played a prominent role in the development of the field of nonlinear dynamics [1]-[3]. The simplicity of its quadratic expression and the richness and intricacy of the properties that stem from it have captivated a large number of scholars and students over decades. It has served as a standard source for the illustration of nonlinear concepts such as: bifurcations, stable and unstable periodic orbits, periodic windows, ergodic and mixing behaviors, chaotic orbits and universality in the sense of the Renormalization Group (RG) method [1]-[3]. It has also become a suitable model system for the exploration of statistical-mechanical structures [4]. All of these properties are shared by one-dimensional unimodal maps where the quadratic maximum is replaced by an extremum of general nonlinearity $z > 1$ [2, 5]. Here we concisely draw attention to the presence of geometrical and scaling laws in the family of attractors generated by the logistic map and to the consequences that these laws have in the dynamical properties of their most interesting object of study: the period-doubling transition to chaos. One-dimensional nonlinear maps, like the logistic, are necessarily dissipative [1] and they settle after much iteration into attractors that may consist of a finite set of points regularly visited or an infinite number of points that may be irregularly (or chaotically) visited [1]-[3]. The logistic map possesses an infinite family of attractors that are connected via cascades of bifurcations at which periods or number of chaotic bands duplicate [1]-[3]. These families of attractors display power laws associated with attractor positions and Pascal Triangles associated with distances between these positions.

We recall that a Pascal Triangle is a triangular arrangement of the binomial coefficients and that it contains many remarkable numerical relations [6]. Blaise Pascal studied this array in the 17th century, although it had been described centuries earlier by the Chinese mathematician Yang Hui, and then by other Indian and Persian scholars. It is therefore known as the Yang Hui triangle in China [7]. This triangle serves as the basis of the De Moivre-Laplace theorem [8] an earlier limited version of the central limit theorem leading to the Gaussian distribution.

We provide below a description of how these properties arise in the logistic map and discuss their implications for the mathematical structures in the dynamics of this nonlinear system, itself a convenient numerical laboratory for the study of statistical-mechanical theories [4].

2. The bifurcation cascades of the logistic map

We briefly recall the basic definitions of the superstable periodic attractors (or supercycles) and the chaotic band-merging attractors (or Misiurewicz points) [1]-[3]. These have become convenient families of attractors in formal descriptions of the bifurcation cascades of unimodal maps, often illustrated by the logistic map $f_\mu(x) = 1 - \mu x^2$, $-1 \leq x \leq 1$, $0 \leq \mu \leq 2$.

The superstable orbits of periods 2^n, $n = 1, 2, 3, \ldots$, are located along the bifurcation forks, i.e. the control parameter value $\mu = \bar{\mu}_n < \mu_\infty$ for the superstable 2^n-attractor is that for which the orbit of period 2^n contains the point $x = 0$, where $\mu_\infty = 1.401155189\ldots$ is the value of μ at the main period-doubling accumulation point. The positions (or phases) of the 2^n-attractor are given by $x_m = f_{\bar{\mu}_n}^{(m)}(0)$, $m = 0, 1, \ldots, 2^n - 1$. The diameters $d_{n,m}$ are defined as $d_{n,m} \equiv x_m - f_{\bar{\mu}_n}^{(2^{n-1})}(x_m)$ [1]. See Fig. 1. Notice that infinitely many other sequences of superstable attractors appear at the period-doubling cascades within the windows of periodic attractors for values of $\mu > \mu_\infty$ [1].

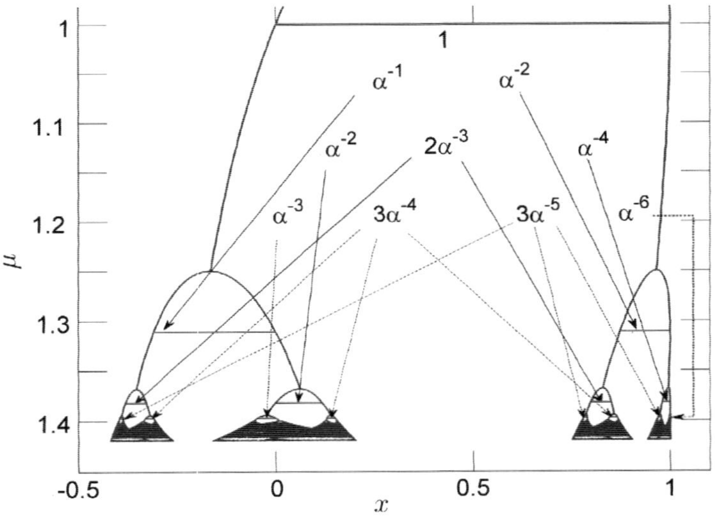

Figure 1. Sector of the period-doubling bifurcation tree for the logistic map $f_\mu(x)$ that shows the formation of a Pascal Triangle of diameter lengths according to the binomial approximation explained in the text, where $\alpha \simeq 2.50291$ is the absolute value of Feigenbaum's universal constant.

When μ is shifted to values larger than μ_∞, $\Delta\mu \equiv \mu - \mu_\infty > 0$, the attractors are chaotic and consist of 2^n bands, $n = 0, 1, 2, \ldots$, where $2^n \sim \Delta\mu^{-\kappa}$, $\kappa = \ln 2 / \ln \delta$, and $\delta = 4.669201609102\ldots$

is the universal constant that measures both the rate of convergence of the values of $\mu = \bar{\mu}_n$ to μ_∞ at period doubling or at band splitting points or Misiurewicz points [2]. The Misiurewicz (M_n) points are attractor merging crises, where multiple pieces of an attractor merge together at the position of an unstable periodic orbit. The M_n points can be determined by evaluation of the trajectories with initial condition $x_0 = 0$ for different values of μ, as these orbits follow the edges of the chaotic bands until at $\mu = \hat{\mu}_n$, the control parameter value for the M_n point, the unstable orbit of period 2^n reaches the merging crises [9]. See Fig. 2. Notice that infinitely many other sequences of Misiurewicz points appear at the band-splitting cascades within the windows of periodic attractors for values of $\mu > \mu_\infty$ [1].

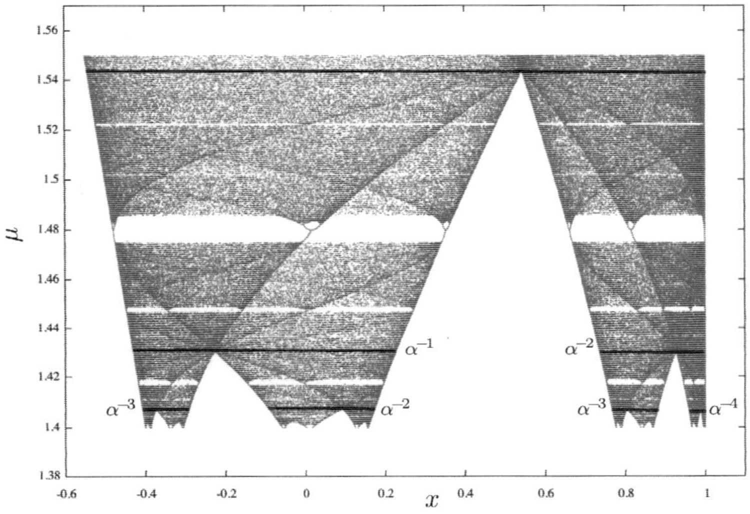

Figure 2. Sector of the main band-splitting cascade for the logistic map $f_\mu(x)$ that shows the formation of a Pascal triangle of band widths (solid lines) at splitting according to the scaling approximation explained in the text, where $\alpha \simeq 2.50291$ is the absolute value of Feigenbaum's universal constant.

3. Pascal Triangles and power-law scaling

The Pascal Triangles and the power-law scaling present in the families of attractors along the bifurcation cascades of the logistic map are best appreciated graphically. These can be visualized in Figs. 1 to 4 that we describe below.

Fig. 1 shows the main period-doubling cascade for the logistic map $f_\mu(x)$ in the (x, μ) plane from period two to its accumulation point (the transition to chaos) and beyond. A few of the supercycle diameters are shown and their approximate sizes are annotated as inverse powers of the (absolute value of the) universal constant $\alpha = -2.50291\ldots$ For convenience we denote $|\alpha|$ as α below and in the figure captions. The diameters naturally assemble into well-defined size groups and the numbers of them in each group can be precisely arranged into a Pascal Triangle. In fact, the diameter lengths within each group are not equal, however the differences in lengths within groups diminishes rapidly as the period 2^n increases [10]. A detailed study of the quantitative differences between the values of the diameters generated by the logistic map and those obtained from the binomial approximation that form the Pascal triangle in Fig. 1 is given in [10]. There are two groups with only one member, the largest and the shortest diameters, and the numbers within each group are given by the binomial coefficients.

Fig. 2 shows a segment of the main band-splitting cascade of the logistic map $f_\mu(x)$ in the (x, μ) plane where we indicate the widths of these bands at the control parameter values $\hat{\mu}_n$ when they each split into two new bands. As in the case of the diameters the band widths diminish in size according to the same inverse powers of α as their numbers 2^n increase. They also form groups of nearly equal sizes with numbers given by the binomial coefficients. Again if it is assumed that for every value of n the widths of comparable lengths have equal lengths then these widths can be obtained from the widths of shortest and longest lengths via a simple scale factor consisting of an inverse power of α. Under this approximation we observe a Pascal Triangle across the band-splitting cascade.

Fig. 3 shows the period-doubling cascade for the logistic map $f_\mu(x)$ in logarithmic scales in order that the power-law scaling of positions of trajectories along the supercycle attractors is plainly observed. These positions are shown as circles. As it can be appreciated in the figure certain sequences of positions each belonging different cycles fall on straight lines that share a common slope. One such alignment corresponds to positions of the principal diameters $d_{n,0}$ that are formed when $x = 0$ is the other endpoint of the interval. (The positions at $x = 0$ do not appear in the figure as their logarithm is minus infinity). The power-law scaling is of the form $|x| \sim \alpha^{-n}$.

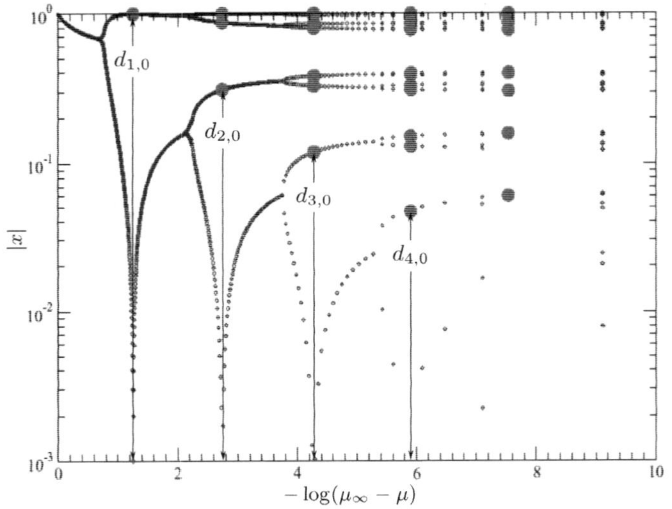

Figure 3. Absolute value of attractor positions for the logistic map $f_\mu(x)$ in logarithmic scale as a function of the logarithm of the control parameter difference $\mu_\infty - \mu$. The supercycle attractors are shown by the arrows and the notation $d_{1,0}, d_{2,0}, \ldots$ corresponds to the so-called principal diameters, the distances between $x = 0$, the position of the maximum of the map, and the nearest cycle position.

Fig. 4 shows the band-doubling cascade of chaotic attractors in the logistic map $f_\mu(x)$ in logarithmic scales so that the power-law scaling of splitting positions M_n is clearly observed. Some of these positions are shown as circles. As it can be observed in the figure sequences of Misiurewicz points fall on straight lines that share a common slope. The power-law scaling is again of the form $|x| \sim \alpha^{-n}$.

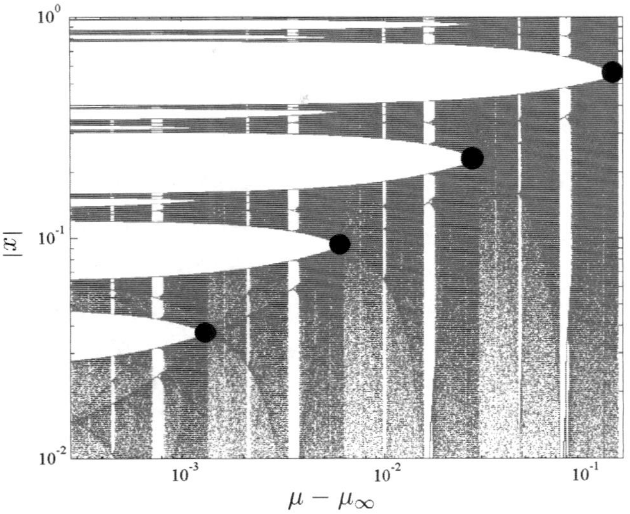

Figure 4. Attractor bands and gaps between them (white horizontal regions) in logarithmic scales, $-\log(\mu - \mu_\infty)$ and $\log(|x|)$ in the horizontal and vertical axes, respectively. The band-splitting points M_n (circles) follow a straight line indicative of power-law scaling. The vertical white strips are periodic attractor windows.

4. Gaussian and multi-scale distributions

As mentioned, the Pascal Triangle is closely associated with a special case of the Central Limit Theorem, known as the De Moivre-Laplace theorem that dates back to 1870 [8]. This theorem establishes that the limiting form of the binomial distribution, the sum of the binomial series of $(p+q)^n$ for which the number of successes s falls between p and q, $p+q=1$, is the Gaussian distribution. Therefore the occurrences of approximate Pascal Triangles for the sets of diameters and bandwidths in the bifurcation cascades implies a Gaussian distribution for these lengths at their accumulation point, the period-doubling onset of chaos. On the other hand, the power-law scaling of supercycle and band-splitting positions is reflected onto the period-doubling onset of chaos as a complete organization of subsequences of positions each following the same power-law scaling. Iteration time evolution at μ_∞ from $t=0$ up to $t \to \infty$ traces the period-doubling cascade progression from $\mu = 0$ up to μ_∞. There is a quantitative relationship between the two developments. Specifically, the trajectory inside the attractor at μ_∞ with initial condition $x_0 = 0$, the 2^∞-supercycle orbit, takes positions x_t, such that the distances between appropriate pairs of them reproduce the diameters $d_{n,m}$, defined for the supercycle orbits with $\bar{\mu}_n < \mu_\infty$ [4]. See Fig. 5, where the absolute value of positions and logarithmic scales are used to illustrate the equivalence.

The limit distributions of sums of positions at the period-doubling transition to chaos in unimodal maps have been studied by making use of the trajectory properties described above [11], [12]. Firstly, the sum of positions as they are visited by a single trajectory within the attractor was found to have a multifractal structure imprinted by that of the accumulation point attractor [11]. It was shown, analytically and numerically, that the sum of values of positions display discrete scale invariance fixed jointly by the universal constant α, and by the period doublings contained in the number of summands. The stationary distribution associated with this sum has a multifractal support given by the period-doubling accumulation point attractor [11]. Secondly, the sum of subsequent positions generated by an ensemble of uniformly distributed initial conditions in the entire phase space was recently determined [12]. It was found that this sum acquires features of the repellor preimage structure that dominates the dynamics toward the attractor. The stationary distribution associated with this ensemble has a hierarchical structure with multi-scale properties [12]. See also Ref. [13].

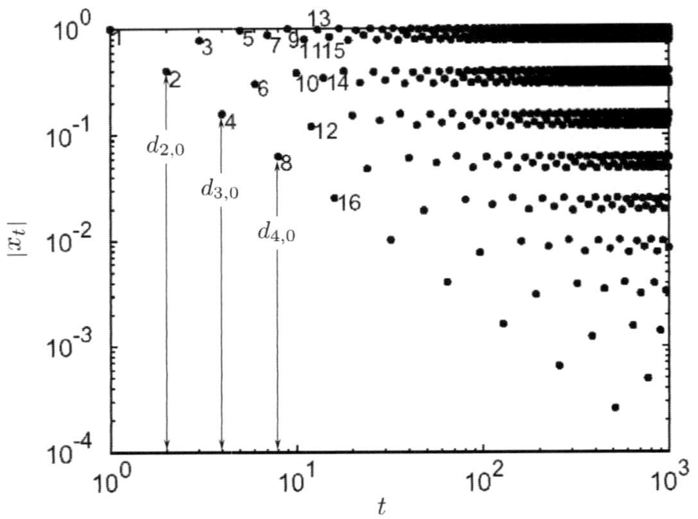

Figure 5. Absolute value of trajectory positions x_t, $t = 0, 1, \ldots$, for the logistic map $f_\mu(x)$ at μ_∞, with initial condition $x_0 = 0$, in logarithmic scale as a function of the logarithm of the time t, also shown by the numbers close to the points.

5. Scale invariant properties and statistical-mechanical structures

Underlying the power-law scaling of positions are the self-affine properties that permeate the dynamics of unimodal maps. As we have seen, these properties manifest visibly along the well-known bifurcation cascades described here. They are also at the center of the Renormalization Group (RG) functional composition developed time ago [1] to provide a firm theoretical basis to the universal properties of the accumulation point of the bifurcation cascades. The RG fixed-point map $f^*(x)$ satisfies the condition $f^*(f^*(x)) = \alpha f^*(x/\alpha)$. Fig. 5 offers us a simple visual opportunity to observe the effect of the RG functional composition and its fixed-point map property. First we notice that certain position subsequences appear aligned in the figure. The most visible is that of $|x_t|$, $t = 2^n$, $n = 0, 1, 2, 3, \ldots$, with a slope equal to $\ln\alpha/\ln 2$, the main diagonal pattern in the figure. Next to it is the subsequence $|x_t|$, $t = 3 \times 2^n$, $n = 0, 1, 2, 3, \ldots$, with the same slope, and so on. All the positions x_t can be distributed into subsequences of the form $|x_t|$, $t = (2l + 1)2^n$, $n = 0, 1, 2, 3, \ldots$, $l = 0, 1, 2, 3, \ldots$, that form a family of lines in the figure with the fixed slope $-\ln\alpha/\ln 2$ [4]. And all of the infinite family of aligned subsequences of positions can be collapsed onto a single line via rescaling with the use of a "waiting" time $t_w = 2l + 1$, $l = 0, 1, 2, 3, \ldots$, a property known as aging in the topic of glassy dynamics [4], [14].

We observe that the positions in the logarithmic scales of Fig. 5 appear grouped into horizontal bands separated by gaps of equal widths. The top band contains one half of the attractor positions as all the odd iteration times appear there. The second band is made of one quarter of the attractor positions, those that correspond to iteration times of the form $t = 2 + 2^n$, $n = 0, 1, 2, 3, \ldots$ And similarly, the $(k + 1)$-th band contains $1/2^n$ of the positions of the attractor, those for iteration times of the form $t = 2^k + 2^n$, $n = 0, 1, 2, 3, \ldots$, $k = 0, 1, 2, 3, \ldots$

The RG method successively transforms the system under study by elimination of degrees of freedom followed by rescaling aimed at restoring its original condition. This procedure can be envisaged in Fig. 5 by eliminating its top band, one half of the total attractor positions, and restoring the original figure by shifting the remaining positions horizontally by an amount $-\ln 2$ and vertically by an amount $\ln \alpha$, in line with the slope $-\ln \alpha / \ln 2$ of the aligned subsequences. This graphical procedure is equivalent to functional composition. Repeated application of this transformation leads asymptotically to the scaling property of the fixed-point map $f^*(x)$.

The scale invariant features of the accumulation point of the bifurcation cascades can be related to statistical-mechanical properties that involve generalized entropy expressions [4], [10].

Acknowledgments

AR is grateful for the hospitality received from the organizers of SPMCS14 in Yichang, China, and acknowledges support from DGAPA-UNAM-IN103814 and CONACyT-CB-2011-167978 (Mexican Agencies). CV acknowledges the support provided by IIMAS-UNAM.

References

[1] Schuster R C 1988 *Deterministic Chaos. An Introduction* (VCH Publishers, Weinheim)
[2] Beck C and Schlogl F 1993 *Thermodynamics of Chaotic Systems* (Cambridge University Press, Cambridge)
[3] Schroeder M 1991 *Fractals, Chaos, Power Laws: Minutes from an Infinite Paradise* (Freeman, New York)
[4] Robledo A 2013 Generalized statistical mechanics at the onset of chaos *Entropy* **15** 5178
[5] van der Weele J P, Capel H W and Kluiving R 1987 Period doubling in maps with a maximum of order z *Physica* **145A** 425
[6] See for example: `http://en.wikipedia.org/wiki/Pascal's_triangle`
[7] See for example: `http://en.wikipedia.org/wiki/Yang_Hui`
[8] See for example: `http://en.wikipedia.org/wiki/De_Moivre-Laplace_theorem`
[9] Grebogi C, Ott E and Yorke J A 1983 Crises, sudden changes in chaotic attractors, and transient chaos *Physica* D **7** 181
[10] Diaz-Ruelas A and Robledo A 2014 Emergent statistical-mechanical structure in the dynamics along the period-doubling route to chaos *Europhys. Lett.* **105** 40004
[11] Fuentes M A and Robledo A 2010 Renormalization group structure for sums of variables generated by incipiently chaotic maps *J. Stat. Mech.* P01001.
[12] Fuentes M A and Robledo A 2014 Sums of variables at the onset of chaos *Eur. Phys. J.* B **87** 32
[13] Diaz-Ruelas A, Fuentes M A and Robledo A 2014 Scaling of distributions of sums of positions for chaotic dynamics at band-splitting points *Europhys. Lett.* **108** 20008
[14] Robledo A 2004 Universal glassy dynamics at noise-perturbed onset of chaos: A route to ergodicity breakdown *Phys. Lett.* A **328** 467

Emergence of diversity in a biological evolution model

R Wang[1], **C Pujos**[1]

[1] Institut Supérieur des Matériaux et Mécaniques Avancés du Mans, 44, Avenue F.A. Bartholdi, 72000 Le Mans, France

E-mail: `wangr@ismans.fr`

Abstract. The ecological systems comprise a rich diversity of species, but the minimum requirements to maintain a large species diversity on long time scales are in general unknown. Here we propose one neutral evolution competition mechanism which is identified to ensure successful conservation of the biodiversity in ecosystems. Here we show that, this mechanism can lead the system into a quasistable state regardless of the initial conditions. By changing one parameter p which represents the probability of introducing one new species, the system can evolve from homogeneous state, the system contains only one species into heterogeneous state, the system contains diversity of species with number of species equal to the system size. This model is simple enough to be analysed theoretically. The theoretical estimation on abundance distribution and the diversity show the same result of simulation. Furthermore, the lifetime of the species is independent of the system size.

1. Introduction

The relationship between biodiversity and ecosystem processes has emerged as a major scientific issue today [1-2]. Multiple species often coexist robustly in natural ecosystems [3]. The interactions between biological species may well be as old as lift itself [4]. The competition, cooperation, or predation are generally considered as majoir determinants for species diversity [5-7]. However it is not easy to keep an ecosystem with diversity when several species compete for the same resources [8]. It has been found that one way to maintain coexistence of multiple species is to include fitness [9]. Although biologists have offered a staggering number of definitions of fitness, they agree broadly on the essence of the idea. In the crudest terms, fitness involves the ability of organisms or, populations or species to survive and reproduce in the environment. The consequence of this survival and reproduction is that organisms contribute genes to the next generation. Another robust way to maintain high diversity is to include space [10-11]. However, Hubbell proposed Unified Neutral Theory of Biodiversity(UNTB) to explain biodiversity patterns without invoking species differences [12-15]. Neutral theory assumes that each individual prospects of death and reproduction are independent of what species it, or its neighbors, belong to. This assumption, while only an approximation, appears to provide a useful description of an ecological community on some spatial and temporal scales. More significantly, it allows the development of a tractable null theory for testing hypotheses about community assembly rules. Based on the neutral theory, here we present an ecosystem model in which species compete within limited system size, and the model is kept as simple as possible to remain analytically tractable.

Journal of Physics: Conference Series **604** (2015) 012019 doi:10.1088/1742-6596/604/1/012019

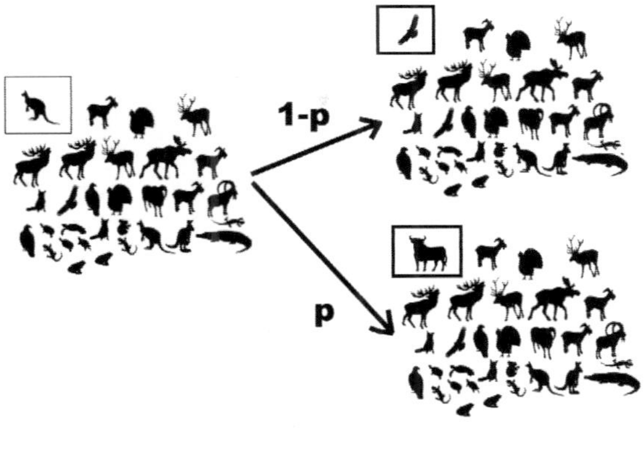

Figure 1. An illustration of the neutral competition model. With probability p, one new species is introduced and replace one individual in the system randomly, while with probability $1 - p$, one randomly chosen individual is duplicated and replaces another in the system randomly.

This paper is devided into four sections. In Sec.II, we introduce an ecosystem model where species compete for resources. In Sec. III, the creation of ecosystem is both computer simulated and analytical calculated, and finally, in Sec.IV, we provide a disscussion of this model.

2. Description of the model

Our model is inspired by the neutral theory, under this assumption, either the species compete with each other equally or being replaced by new species equally within the constant system size. Specificly, we consider a system with fixed population N. Initially, there are M species, and each species with N/M individuals. We then study by computer simulation. At each step of the simulation, we either introduce a single new species and replace it with an old randomly chosen individual, or we randomly chose two individuals and replace one by another. The detail computer simulation are, at each step we chose one individual i, $(i = 1, 2, ..N)$ despite its species i_s, $(s = 1, 2, ...M)$ at random, and do the following(see fig.1).

(1) With probability p, we introduce one individual k which belongs to a new species k_s, and replace i with k.

(2) Otherwise (i.e., with probability $1 - p$) we randomly pick a individual j, and replace it by i.

Step (1) represents the introduction of new species into teh system. In order to keep the total population as a constant, one old individual must be replaced by the new one. Step (2) represents the neutral competition of the individuals in the system. During this dynamics process, if the last individual of one species is died, this species' lifetime is the time from its born until the last single one dies. Note that total population of system N is fixed, and each individual can be selected equally regardless of its species.

3. Results

The expected qualitative behavior of the model is clear. Since when $p = 0$, no new species can be introduced, the species in the system exclude each other until all of the individuals belongs to the same type. In this limite, the diversity D (the number of species) is 1. While in the other extreme situation, when $p = 1$, one new species is introduced at each time, the system evolves into the state in which each individual belongs to different species, thus, diversity is equal to

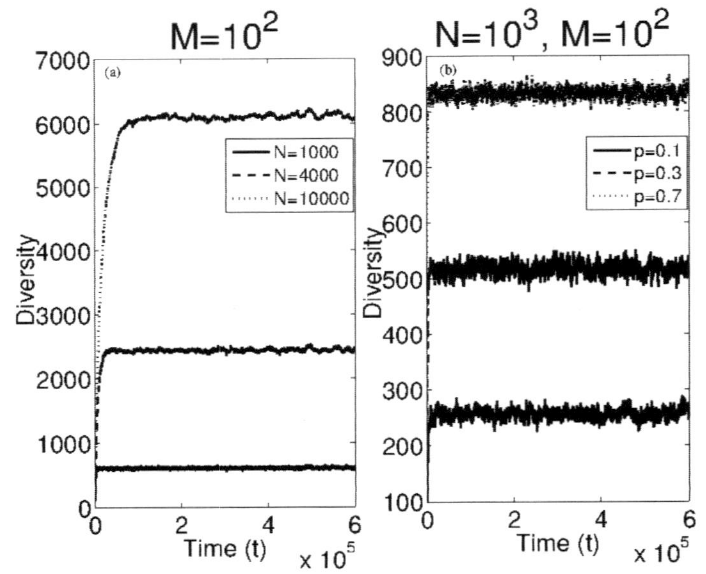

Figure 2. Evolution of diversity for (a) different system size $N(N = 1000,\ 4000{,}10000)$ with the number of species $M = 100$, when $p = 0.4$ and (b) different probability $p(p = 0.1,\ 0.3,\ 0.7)$ with system size $N = 1000$, and number of species $M = 100$.

system size N. When p is between 0 and 1, the system will reach to a quasistable state with fluctuation as shown in Fig.2. Here we start with constant initial number of species $M = 100$ with different system size N.

It is observed from Fig.2 that the system evolves to a quasistable state, in which diversity fluctuates around its average value. In the final stable state, the average diversity increases with increasing system size N and p, and thus, species abundance also becomes stable which is due to the conservation of the total population during the evolution.

Due to the neutral replacing and duplicating mechnamics, this model can be theoretically analyzed. Considering the birth and replace process in this system, at each time, any species can only increase one or decrease one individual, i.e., the number of species with certain abundance can only increase one or decrease one. In a word, the "species flow"–the number of species which changed their abundance-must be conserved: the total species flow into one andance should be equal to the total species flow out of it (as shown in Fig.3). Let $f_i(i = 1, 2, ...N)$ represent the number of species with abundance i, we have the simultaneous equations as following when taking into account the conservation of species flow:

$$\begin{cases} f_1\frac{1}{N}[p + (1-p)(1-\frac{1}{N})] = p \\ f_2\frac{2}{N}[p + (1-p)(1-\frac{2}{N})] = f_1\frac{1}{N}(1-\frac{1}{N})(1-p) \\ \cdots \\ f_k\frac{k}{N}[p + (1-p)(1-\frac{k}{N})] = f_{k-1}\frac{k-1}{N}(1-\frac{k-1}{N})(1-p) \\ \cdots \\ f_{N-1}\frac{N-1}{N}[p + (1-p)(1-\frac{N-1}{N})] = f_{N-2}\frac{N-2}{N}(1-\frac{N-2}{N})(1-p) \\ f_N\frac{N}{N}[p] = f_{N-1}\frac{N-1}{N}(1-\frac{N-1}{N})(1-p) \end{cases} \tag{1}$$

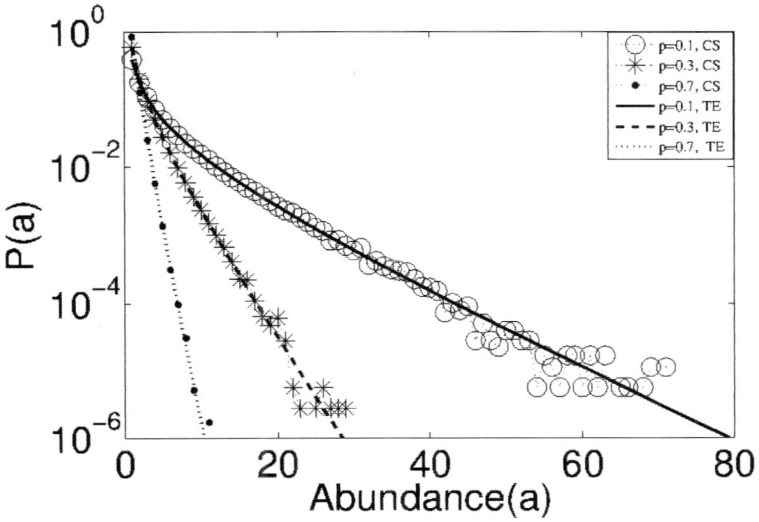

Figure 3. Compter simulation result(CS) and theoretical estimation(TE) result for abundance distribution when $p = 0.1$, $p = 0.4$ and $p = 0.7$ with system size $N = 1000$.

The first row of Eq(1) represents the conservation of species flow for species with abundance equal to 1. The right hand side is the total inflow, which is the probability p of only one single new species, while the left hand side consists two terms which contribute to the total outflow. The first term corresponds to dynamics step (1) that the species which has only single individual is chosen and its individual is replaced by one new born species. The second term accounts for dynamics step (2) that those single individual composed species is chosen and its individual is replaced by the other species in this system. And so forth for the other rows which corresponds to each species abundance.

The result for abundance $f_k(k = 1, 2, ..., N)$ could expresses as:

$$f_k = \frac{Np}{k}\frac{N!}{(N-k)!}\frac{(1-p)^{k-1}}{\prod_{i=N-k}^{N-1}[Np + i(1-p)]} \tag{2}$$

From Eq.(2), it is obvious that the abundance distribution is exponential function under the system size limit($N \to \infty$):

$$f_k \sim p(1-p)^{k-1} \tag{3}$$

Both the computer simulation(CS) result and the theoretical estimation(TE) result for abundance distribution are shown in Fig.3. It is found that the system abundance decreases with increasing p, which can be explained as following. When $p = 0$, the system will end up with one species, which means abundance is N. On the contrary, when $p = 1$, there will be only one sigle individual for each species in the system, which leads to abundance equal to 1. Thus, as p between 0 and 1, the larger p, the more new species are introduced into system, the smaller of the species abundance.

From the "species flow"deduced abundance expression Eq.(2), it is easy to calculate the

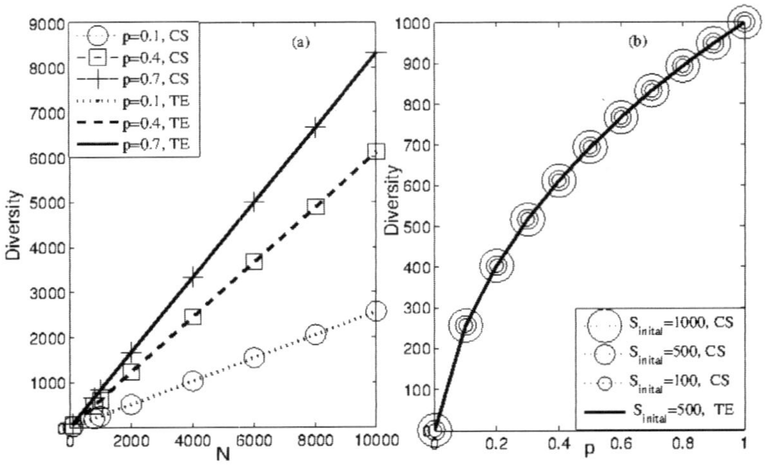

Figure 4. Computer simulation result(CS) and theoretical estimation(TE) result for diversity as a function of (a) system size N under different $p = 0.1$, $p = 0.4$, $p = 0.7$ and (b) p with different initial species $S_{inital} = 1000$, $S_{inital} = 500$,$S_{inital} = 100$.

diversity which is the total sum of species with different abundance. It expresses as:

$$D = \sum_{k=1}^{N} f_k = \sum_{k=1}^{N} \frac{Np}{k} \frac{N!}{(N-k)!} \frac{(1-p)^{k-1}}{\prod_{i=N-k}^{N-1}[Np + i(1-p)]} \tag{4}$$

The equation (4) is approximately equal to $\frac{Np}{1-p}E_i$ with $E_i = \sum_{k=1}^{N} \frac{(1-p)^k}{k}$. It is found that diversity is a linear function of system size N, and nonlinear function of p. The result for both computer simulation and theoretical estimation are shown in Fig.4. As stated before, the larger p, the larger diversity, and the larger system size N leads to larger diversity. It is found that the final steady state is independent of the initial number of species, it can be understood that for the competitive exclusion process, each individual is equally chosen despite its species,which means all types behave symmetrically, as a result, the initial number of species have no influence on the system steady state.

The dynamics of the population of a given species is governed by generalized birth and death events. Let $p(n,t)$ denote the probability that one species contains n individuals at time t. In this simplest scenario, the time evolution of $p(n,t)$ is regulated by the following equation:

$$\begin{aligned} p(n,t+1) &= p(n-1,t)\frac{N-(n-1)}{N}\frac{n-1}{N}(1-p) \\ &+p(n,t)\frac{N-n}{N}[\frac{N-n}{N}(1-p)+p] + \frac{n}{N}\frac{n}{N}(1-p) \\ &+p(n+1,t)\frac{n+1}{N}[\frac{N-(n+1)}{N}(1-p)+p] \end{aligned} \tag{5}$$

The right hand side is composed by three terms. The first term is the probability that species increase its individuals from $n-1$ to n by replacing other type of individuals, and the second is

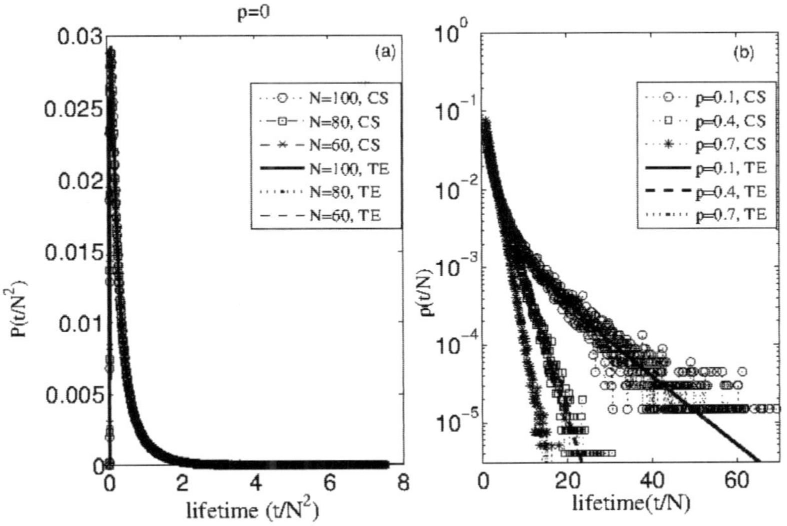

Figure 5. Computer simulation result(CS) and the theoretical estimation(TE) result for rescaled lifetime distribution as (a)$p = 0$ and (b) $p = 0.1, 0.4$ and 0.7

the probability that the species remain the same number of individuals, either by introdcing a new single species which replacs other types of species or by the duplicated individual replacing its own type,and the last term is the probability that species decrease one individual from $n+1$ to n by either introducing a new single species or duplicating another species. This equation is the finite difference partial differential equation, which can only be solved numerically. Both the simulation result and numerical equation results are shown in Fig.5. The lifetime is a log-normal distribution when $p = 0$, and can be rescaled by system size. The Fig.5 (b) shows that the lifetime can be rescaled by system size N for $p > 0$, and become shorter when p is larger. This is easy to understand,the more frequently new species is introduced to this system, the more individuals will be replaced, which leads to shorter lifetime, and is proportional to system size N.

4. Discussion

Here we proposed one neutral competition evolution model to explain the biodiversity in ecological system. This simple model is controlled by one paramater p, which represents the probability of introducing one new species. Under constant system size, the diversity increases to systems size N when $p = 1$, or decreases to 1 when $p = 0$, or a quasistable state with fluctuation ($0 < p < 1$) when system reaches to steady state. By theoretically analyzing the "species flow"equations, it is found that the diversity grows linearly with systems size N, and nonlinear with p. The abundance is exponential distribution and decreases with increasing p. Finally, the lifetime can be rescaled by system size. It is log-normal distribution when $p = 0$ and decreases with increasing p when $p > 0$. Furthermore, this neutral competition model is simple enough to make theoretical analysis. The theoretical estimation gives the same results as the computer simulation, which provide one framework for the more complicated situation investigation in future work.

References

[1] Eigen M and Schuster P 1979 *The Hypercycle: A Principle of Natural Self-Organization* (Berline: Springer).
[2] Hooper D U *et al.* 2005 *Ecol Monogr* **75** 3
[3] Ives A R and Carpenter 2007 *Science* **317** 58
[4] Wootton J T and Emmerson M 2005 *Annu Rev Ecol Evol Syst* **36** 419
[5] Robert May M 1973 *Stability and Complexity in Model Ecosystems (MPB-6) (Monographs in Population Biology)* (New Jersey: Princeton University Press)
[6] Lidicker W Z 1979 *BioScience* **29** 475
[7] Luca P, Alberto B, Seven E J 2013 *Ecological Processes Handbook* (CRC Press)
[8] Hardin G 1960 *Science* **131** 1292
[9] Reed D H and Frankham R 2003 *Conserv. Biol.* **17** 230
[10] Hufkens K, Scheunders P and Ceulemans 2009 *Ecol. Res.* **24** 977
[11] Kerr B, Neuhauser, Bohannan B J M and Dean A M 2006 *Nature* **442** 75
[12] Hubbel S P 1997 *Coral Reefs* **16** 9
[13] Hubbel S P 2001 *Unified Neutral Theory of Biodiversity and Biogeography* (Princeton:Princeton University Press)
[14] Chave J 2004 *Ecol. Lett.* **7** 241
[15] Volkov I, Banavar J R, Hubbel S P and Maritan A 2003 *Nature* 424 1035

Emergence of moving pattern in a collective game

Han Fu, Yongbin Shi and Yougui Wang[1]

School of Systems Science, Beijing Normal University, Beijing 100875, P.R. China

E-mail: ygwang@bnu.edu.cn

Abstract. In this paper, we propose a collective game where agents, distributed in one-dimensional space, seek for their best positions through adaptive learning. The movements of agents are driven by a simple rule: they are attracted by others with higher status and yet restricted by their initial positions, according to which a benefit function is formulated. It is found that some fascinating complex patterns arise from such simple rules. Aggregations of agents emerge in the form of clusters along this position line after a dozen rounds. While agents strive to move forward, a chain of clusters shift backward and ultimately collapse one by one at an almost fixed point. The trail of shifting behaves like an attractor which attracts different groups of agents to join in. Such intricate moving pattern turns out quite sharp and clear when agents have a uniform pace length.

1. Introduction

It is commonly believed that complexity originates in simplicity. One of the most fundamental principles of Stephen Wolfram's *A New Kind of Science* lies in the insistence that more complex rules do not necessarily lead to greater complexity in overall behaviour. In other word, simple programs with straight forward rules, such as elementary cellular automata, are sufficient to capture the essence of almost any complex process [1]. Wolfram's research as well as other pioneers' endeavours such as John Conway's Game of Life [2] reveal the computational nature of social sciences and promote the development of the discipline known as "computational social science".

Parallel with cellular automata, agent-based modeling (ABM) is another effective means to cope with problems of sociology and the complexity of social dynamics [3-5]. With a bottom-up approach, agent-based models are a kind of microscopic model which attempts to investigate and sometimes predict agents' collective behaviour, broadly known as occurrence of emergence [6]. Within the last decades, ABM has expanded in social science studies and scored remarkable achievements, including some renowned models such as *Boids* simulating the flocking behaviour of birds [7] and *Sugarscape* exploring social phenomena like seasonal migrations [8].

A focus issue within social sciences never fails to fascinate researchers: the causes of aggregation. Aggregation is ubiquitous in nature, such as a flock of birds or a school of fish. In human society, it is also a common social phenomenon that people with different social status, political power, and religious belief are inclined to form separate communities, broadly referred to as social stratification [9]. Substantial endeavours have been made to search for explanatory insight into the formation. There is research verifying that aggregation results from collaborative interactions [10,11] and human evolved psychological processes [12].

[1] To whom any correspondence should be addressed.

On the basis of previous studies, we regard an agent-based model with simple interaction rules and adaptive learning players as a valid approach to understanding aggregate phenomena. Such a model was proposed in our recent research where agents distributed in one dimensional space are driven by benefits obtained from higher-status ones, the static pattern of which has exhibited a fractal characteristic [13]. To simulate human society, the social climbing phenomenon in which people prefer to forge links with others with higher social status has attracted attention [14,15]. In our previous work we assumed that people are not only catching up with others with higher status and yet restricted by their initial positions. In this work, we further investigate such a simple model and astonishingly unearth even more complicated and enchanting dynamic processes.

The remaining parts of this paper are organized as follows. Section 2 of this paper elaborates on how the model runs. An overall evolution of agents' location distribution over time is displayed. In-depth analysis of the formation of moving patterns is offered in Section 3. The short-term and long-term movements of a group of agents who aggregate at a specific time point are examined. The moving trails of all agents, grouped by pace length and initial position, are exhibited. A particular case is discussed where pace lengths are unified. In Section 4 we conclude.

2. Model description

To explore the generation mechanism of aggregations, we design such an agent-based model where agents locate on a line and adjust their locations to seek for the maximum benefits. The interacting rules of the model and some preliminary results are presented in this section.

2.1. Interacting rules

At the very beginning, each agent is endowed with an initial position randomly distributed in this one-dimensional space of finite length. Then the position adjustments are made at each time step by every agent in order to maximize his benefit. The benefit function for one agent involves his current position and those of other agents with higher status, as well as his initial position. Specifically, the benefit gained by agent i from agent j at time t is formulated as follows,

$$\begin{cases} 0, & if\ x_j(t) < x_i(t); \\ \frac{1}{2}\left[x_i(t) + x_j(t)\right] - x_i(0), & if\ x_j(t) \geq x_i(t), \end{cases} \tag{1}$$

where $x_i(t)$ represents agent i's position at time t, and hence $x_i(0)$ is his initial position. Only people ahead make contributions to his benefit, whereas those behind do not. By summing up all contributions from others, we express expected benefit of agent i in the form of the distribution of all agents' current positions as well as his own initial position:

$$B_i(t) = \sum_{j \neq i}^{N} B_{ij}(t). \tag{2}$$

Since agent i is assumed to be an adaptive actor, he has only three choices of next move: staying at current position, moving upward and moving downward. There is always a trade-off when he makes his choice of movement. Suppose he moves forward, the higher position will bring him more benefits but the number of agents ahead would be reduced and his benefit will then be lessened. He makes his decision based on the comparison of the respective benefits induced by the three options. The shift in location is given by

$$x_i(t+1) = x_i(t) + sign \cdot \delta_i, \tag{3}$$

where $sign = -1, 0, 1$ and δ_i denotes the pace length of agent i. The agent makes the decision by calculating and comparing the benefits of these three possible positions at time $t + 1$ while expecting others to retain their positions at time t. Each move always guarantees the agent the maximum benefit at the current time point.

2.2. Simulation results

In the beginning, 500 agents are positioned randomly from 0 to 1 on a number axis, and they will be restricted within this interval hereafter. Then, they will adjust their locations to seek for higher benefit.

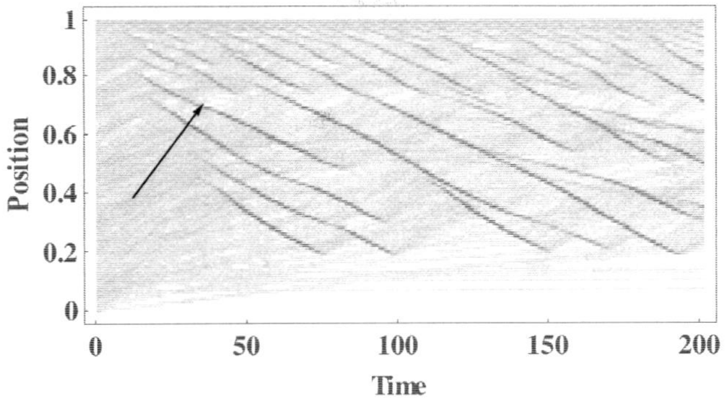

Figure 1. Evolution of the distribution of agents' positions over time, with the horizontal and vertical directions representing time and space respectively.

Their moving abilities are set to be diverse and each agent is endowed with a certain pace length which is randomly sampled from a uniform distribution from 0 to 0.01. Given these parameters, we ran the model and recorded the evolution of the locations of all agents over a period of 200 time steps.

A visual representation of how the distribution of agents along the position line evolves with time is shown in figure 1, where horizontal and vertical axes represent time and space respectively. The colours in this figure specify the corresponding densities of population – a darker pixel corresponds to a higher density here and now.

When the simulation started, the agents are distributed randomly in the space, so the colours at the beginning are light and nearly even. After about 20 rounds, darker pixels come into being which means several clusters of agents emerge. The right-downward dark stripes can be obviously observed in this figure, and they disappear at almost a same position around 0.2. In other word, the clusters shift downward as time goes by and ultimately collapse at one point.

3. Formation of moving patterns

3.1. Moving tracks of a group of agents

In order to further analyze the complicated pattern arising from interaction among a group of benefit seekers, whose ability of movement and initial position are different in our simulation settings, we

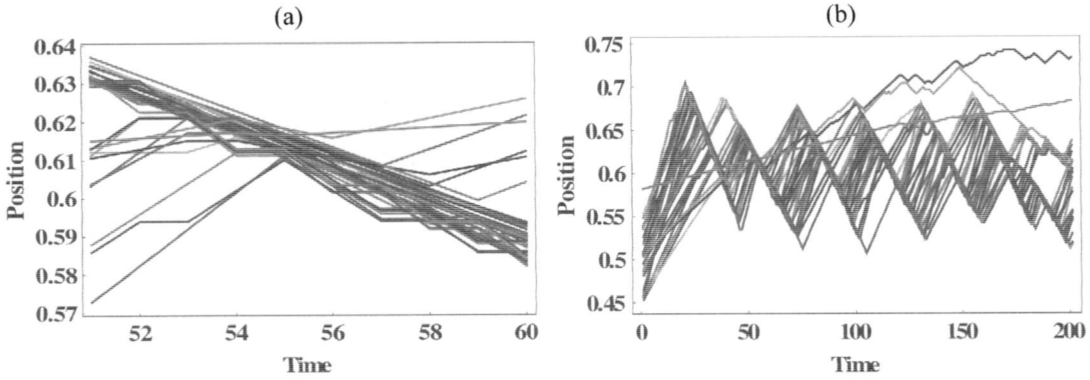

Figure 2. Moving tracks of agents who are included in this cluster at time 55 in the short run (a) and the long run (b).

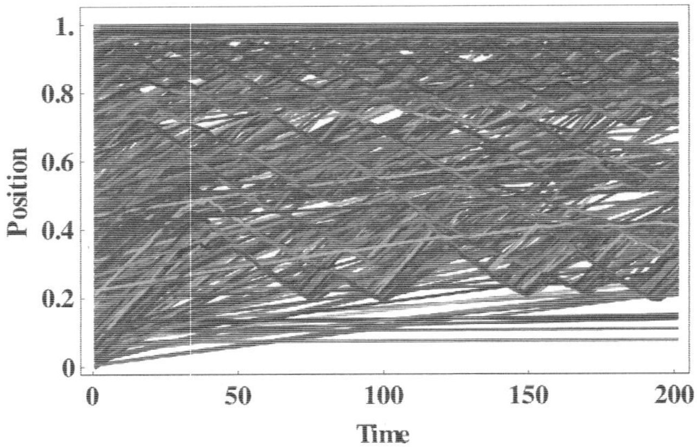

Figure 3. Moving tracks of all agents during the whole process.

carried out the following analysis. Firstly, we narrow our focus to one single cluster by observing movement of agents who aggregate to form a cluster at a specific time point. To be more specific, we choose to investigate the cluster indicated by the black arrow in figure 1. We trace the movement of the agents who are included in this cluster at time 55 in the short term and long term.

Figure 2(a) plots the movement trails of this group of agents in the short run (from time 51 to time 60) and each curve corresponds to one individual agent. It is observed that some of the agents keep pace with each other and remain in this cluster throughout the short period, while others just pass by and leave the cluster after the short encounter. Looking into the trails of this group during the whole process, we can get more interesting findings, as shown in figure 2(b). In the long term, most agents in this group move upward and downward in coordination with each other. In most cases, they go upward successively, but downward synchronously, this is the reason for the formation of the clusters. One possible explanation is that once the agent with the highest status in this group moves downward, he would mobilize more agents to follow immediately for fear of benefit loss.

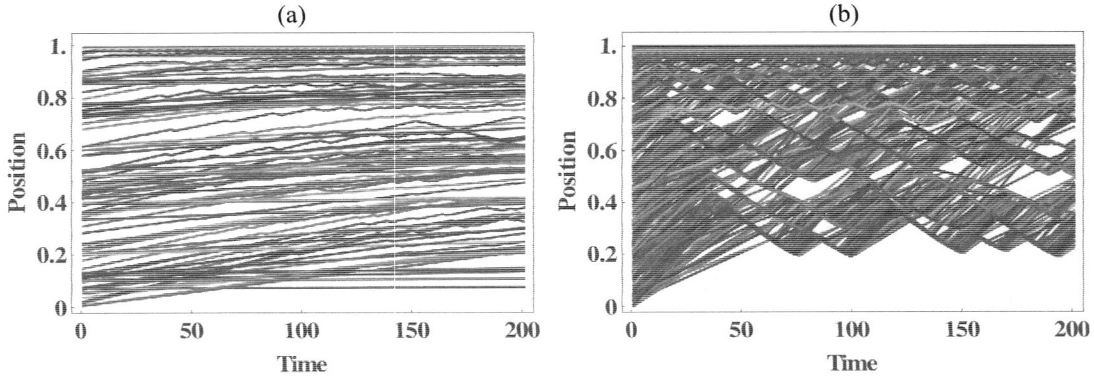

Figure 4. Moving tracks of agents with smaller pace lengths (a) and larger pace lengths (b).

3.2. Moving tracks of all agents

We can explore all agents' movements by plotting the changes in positions of all agents during the whole process in figure 3. It is found that it is not the same group of agents who accomplish the shift of a cluster mentioned above, but a combination of several different groups of agents, who move around different centre points. It appears that different groups are playing a relay game: one group of people relay the downward-shifting cluster to the next group and then move upward themselves, resulting in those amazing dark stripes in figure 1. It is difficult to interpret that the shifts of clusters behave like an attractor which can connect several downward segments perfectly to form a completed path.

By selecting those with little pace lengths (less than 0.002 here) (see figure 4(a)), it is found that these people are hardly influenced by others due to their limited moving capacities. They merely keep moving forward at their own speed all the time. By selecting those with larger pace lengths (more than 0.003 here) (see figure 4(b)), the structure of the aggregation pattern is presented more clearly.

To examine the distinctions among movements of agents with different levels of initial positions, we divide all agents into three groups, one with lower initial positions (from 0 to 0.33), one with medium initial positions (from 0.33 to 0.66), and another one with higher initial positions (from 0.66 to 1). In figure 5, movements of the three groups are displayed respectively. Almost every agent will finally improve his status by moving upward, but he cannot improve it significantly due to the limitation of initial status. Some of lower-level agents break the bonds and intrude into the medium area; a smaller proportion of medium-level agents promote their status to achieve the high level; none of the agents with initial higher status ever leaves the "high zone".

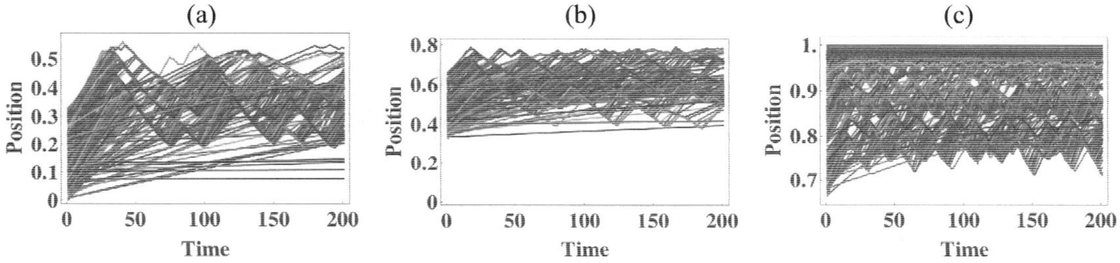

Figure 5. Moving tracks of agents with (a) lower initial positions (from 0 to 0.33); (b) medium initial positions (from 0.33 to 0.66); (c) higher initial positions (from 0.66 to 1).

3.3. *Uniform pace length*

In the particular case where all agents have the same pace length (0.005 here), which means the sole randomness remained comes from the random initial positions, the intricate aggregation pattern still appears and displays a clearer structure, as shown in figure 6. The relays of shifting clusters constitute a beautiful carpet pattern which is both intriguing and bewildering. What factors prompt the groups of ascending agents to begin to turn back and the descending ones to begin to gain higher status remain ambiguous to us.

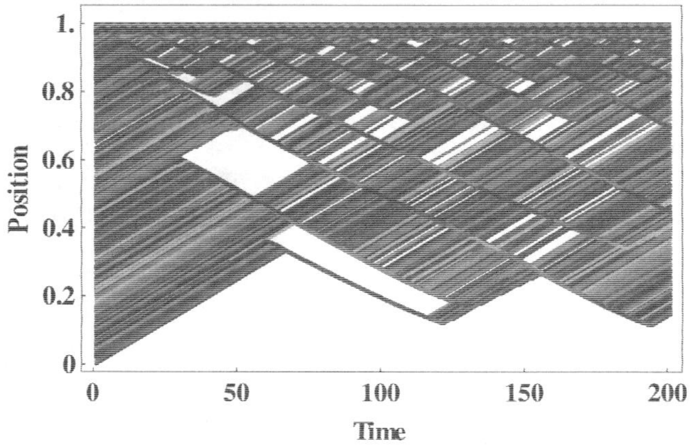

Figure 6. Moving tracks of agents with uniform pace length.

4. Conclusion

Complexity arises from simple rules, which is a common belief in complexity science. Recently, W. Brian Arthur argued that "an economy that is not dead, static, timeless, and perfect, but one that is alive, ever-changing, organic, and full of messy vitality" [16]. In this paper, a vivid demonstration of this conception is exhibited in a collective best-position searching model. The agents adaptively adjust their positions in order to get more benefit. The benefit for one agent obtained from others with higher status is formulated by the sum of differences between the mean value of his and another one's current positions, and his initial position. The simulation results of the model show that the agents will aggregate as clusters when they are moving upward separately for larger benefits. At first glance, the clusters are generated by synchronized downward move in one group. Actually, from an overall perspective, the clusters shift in the form of a "relay race" accomplished by different groups of people successively. The moving aggregation pattern becomes more distinct when we assume all agents possess the same movement ability. However, how this complex pattern is derived from those simple rules remains ambiguous and incomprehensible so far.

This model provides a perspective to understand the formation mechanism of some social norms, that is to say, individual behaviour of each agent is motivated by both his inner psychological desire and the social circumstance or situation he is facing, while the circumstances are co-created by all individuals. In the course of interpersonal interaction, people form spontaneous communities within the whole society since collective actions undoubtedly offer protection and benefit. Following this perspective, we believe that many other complicated sociological phenomena can be understood by taking beneficial interaction and mobility of agents into account.

References

[1] Wolfram S 2002 *A new kind of science* (Champaign: Wolfram media)
[2] Gardner M 1970 Mathematical games: The fantastic combinations of John Conway's new solitaire game "life" *Scientific American* **223(4)** 120-123
[3] Gilbert N and Terna P 2000 How to build and use agent-based models in social science *Mind & Society* **1(1)**57-72
[4] Tesfatsion L 2003 Agent-based computational economics: modeling economies as complex adaptive systems *Information Sciences* **149(4)**262-268
[5] Epstein J M 1999 Agent-based computational models and generative social science *Generative Social Science: Studies in Agent-Based Computational Modeling* **4(5)**4-46
[6] Gilbert N 2008 *Agent-based models* (Sage)
[7] Reynolds C W 1987 Flocks, herds and schools: A distributed behavioral model *ACM SIGGRAPH Computer Graphics* **21(4)**25-34
[8] Epstein J M 1996 *Growing artificial societies: social science from the bottom up* (Brookings Institution Press)
[9] Kerbo H 1996 *Social stratification* (New York: McGrawHill)
[10] Parrish J Kand and Hamner W M (Eds.) 1997 *Animal groups in three dimensions: how species aggregate* (New York: Cambridge University Press)
[11] Turchin P 1998 *Quantitative analysis of movement: measuring and modeling population redistribution in animals and plants (Vol. 1)* (Sunderland, Massachusetts, USA: Sinauer Associates)
[12] Barkow J H 1992 Beneath new culture is old psychology: Gossip and social stratification. In Barkow J, Cosmides L and Tooby J (eds.) *The Adapted Mind: Evolutionary Psychology and the Generation of Culture* (Oxford University Press) 627-637
[13] Wang C, Xiong W T, and Wang Y G 2012 Self-Similarity in Game-Locked Aggregation *Chinese Physics Letters* **29(12)**128903
[14] Bhattacharya H and Dugar S 2014 Partnership Formation: The Role of Social Status *Management Science* **60(5)**1130-1147
[15] Xie W J, Li M X, Jiang Z Q, and Zhou W X 2014 Triadic motifs in the dependence networks of

virtual societies *Scientific Reports* **4**, 5244
[16] Arthur W B 2015 Complexity economics: a different framework for economic thought, in Arthur W B (eds.) *Complexity and the Economy* (New York: Oxford University Press)

Dynamics of self-similar pulse generated in normal dispersion-decreasing fiber

Ge Xia [1], **Youceng Feng** [2]
[1] School of Electronic and Electrical Engineering, Wuhan Textile University, Wuhan, 430073, China
[2] College of Electronics Information Engineering, South-central University for Nationalities, Wuhan, 430074, China

E-mail: xiaboat@wtu.edu.cn, fengyc@mail.scuec.edu.cn

Abstract. Self-similar pulse can be generated in a normal dispersion-decreasing fiber. We show that the dynamics of self-similar pulse evolution is governed by the different spectrum broadening mechanisms in the different stages, which is confirmed by the typical pulse spectral structures with amplitude noise added.

1. Introduction

Self-similar solutions to the nonlinear equations have been a subject of close attention in various physical systems [1]. Self-similar evolution of short pulses propagating in a fiber is of particular interest in fiber optics [2]. Mathematically, the exact and asymptotically stable self-similarly evolving solutions are governed by the generalized nonlinear Schrödinger equation (NLSE) describing propagation dynamics of short pulses in the fiber [3]. Specifically, it was demonstrated that the pulse with an arbitrary shape after propagation in a passive fiber with decreasing normal dispersion transforms into an asymptotically parabolic self-similar pulse [4], and can be used for pulse compression and supercontinuum [5]. However, optical wave breaking (WB) may occur during the self-similar pulse evolution in a fiber with normal dispersion under highly nonlinear condition for a given picosecond seed pulse [6].

In this work, we show that self-similar pulse evolution in a normal dispersion-decreasing fiber can be characterized by a dimensionless pulse shape function in the frequency domain as well as in the time domain. In particular, the dynamics of pulse evolution is governed by the different spectrum broadening mechanisms in the different stages once WB takes place, which is confirmed by the typical pulse spectral structures with amplitude noise added.

2. Self-similar pulse evolution in normal dispersion-decreasing fiber

The propagation of short optical pulses in a normal dispersion-decreasing fiber (NDDF) with varying group-velocity dispersion (GVD) may be described in terms of the dimensionless NLSE with varying coefficients

$$i\frac{\partial q}{\partial z}-\frac{\beta_2}{2}D(z)\frac{\partial^2 q}{\partial t^2}+\gamma|q|^2 q=0 \quad (1)$$

Where the $q(z,t)$ is the complex amplitude of the slowly varying optical field, $\beta_2>0$ is the GVD value at $z=0$, and γ is the nonlinearity coefficient, which is considered to be uniform along NDDF, and higher-order dispersion is not included in Equation (1) while it is possible to fabricate NDDF with third-order dispersion by introducing a W-shaped core according to [4]. The normalized function $1 \geq D(z) > 0$ describes the variation of the fiber GVD along its length. Specifically, with the choice of the hyperbolic profile of dispersion

$$D(z) = \frac{1}{(1+\Gamma_0 z)} \tag{2}$$

Equation (1) can be solved by using numerical approach of the symmetrical split step Fourier method with all the pulse and fiber parameters are chosen from [4]: A Gaussian chirp-free input pulse with full width half maximum (FWHM) 1.0ps and pulse energy of 40pJ, the fiber parameters: $\gamma = 3.33\text{km}^{-1}\text{W}^{-1}$, $\beta_2 = 1.25\text{ps}^2.\text{km}^{-1}$, $\Gamma_0 = 0.028\text{m}^{-1}$. From these parameters, one obtains dispersion length $L_D = 288\text{m}$ and nonlinearity length $L_{NL} = 8\text{m}$, $N^2 = L_D/L_{NL} = 36 >> 1$, indicating a highly nonlinear condition for the given picosecond seed pulse propagating in the fiber. Figure 1(a) and Figure 1(b) show the pulse temporal and spectral evolution along the fiber at different lengths.

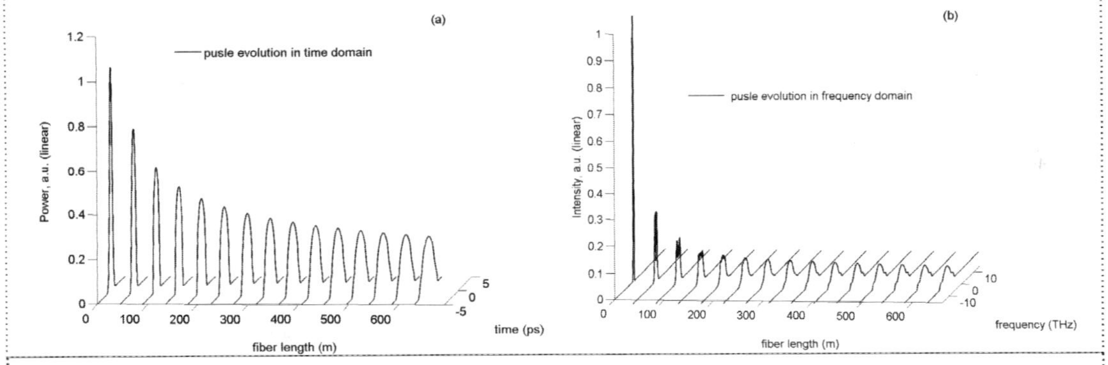

Figure 1. The self-similar pulse evolution along the fiber in the time and frequency domain.

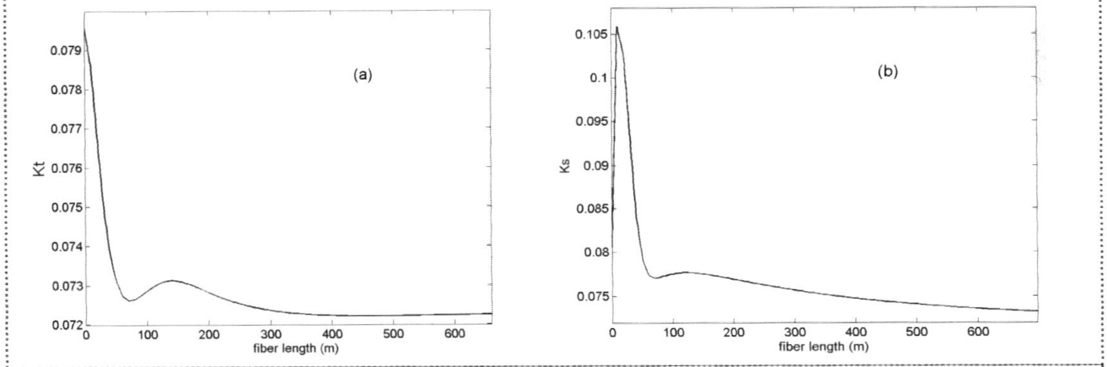

Figure 2. The evolution curves of *K*t and *K*s along the fiber in the time and frequency domain.

To describe the pulse evolution quantitatively, a dimensionless function *K*t is usually introduced to characterize the pulse temporal shape, defined as

$$Kt\left[q(z,t)\right] = \frac{\int_{-\infty}^{\infty} t^2 \left|q(z,t)\right|^2 dt \cdot \left(\int_{-\infty}^{\infty} \left|q(z,t)\right|^4 dt\right)^2}{\left(\int_{-\infty}^{\infty} \left|q(z,t)\right|^2 dt\right)^5} \tag{3}$$

*K*t has been associated with an invariant of the propagation equations of the second-order moments of the pulse amplitude and effective power within the parabolic approximation for the pulse phase. This invariant indirectly expresses the conservation of energy [7]. It has been pointed out that $K = 0.0796$ for a Gaussian pulse and $K = 0.0720$ for an ideal parabolic pulse. The closer *K* approaches 0.0720, the better pulses evolve to self-similar parabolic pulses. Here we further introduce it into the frequency domain as follows

$$Ks\left[s(z,\omega)\right] = \frac{\int_{-\infty}^{\infty} \omega^2 \left|s(z,\omega)\right|^2 \mathrm{d}\omega \cdot \left(\int_{-\infty}^{\infty} \left|s(z,\omega)\right|^4 \mathrm{d}\omega\right)^2}{\left(\int_{-\infty}^{\infty} \left|s(z,\omega)\right|^2 \mathrm{d}\omega\right)^5} \tag{4}$$

Where $s(z,\omega)$ denotes the pulse spectrum envelope, and it is the Fourier transformation of $q(z,t)$. The validity of Equation (4) is guaranteed by the Energy conservation law.

The evolution curves of *K*t and *K*s versus the pulse transmission distance are illustrated in Figure 2(a) and Figure 2(b) respectively. From the Figures, we can see both curves approach asymptotically towards the 0.0720 level line during pulse evolution though *K*s curve fluctuates more severely than *K*t curve, which indicates that the pulse will gradually evolve into a self-similar parabolic shape pulse eventually if the pulse propagates a long enough fiber length (far above 650m).

3. Dynamics of self-similar pulse evolution

As it can be seen in Figure 2(a) and Figure 2 (b), both curves have an obvious oscillation before it come into the later steady self-similar evolution, which reveals that WB starts to take place. It is known that WB results from the temporal overtaking of different parts of pulse due to their different velocity forward in the pulse caused by the nonmonotonic pulse chirp. When the shifted light overruns the pulse tails, the leading and trailing regions of the pulse then contain light at two different frequencies, which interfere and generate new frequencies. The underlying mechanism of this kind of pulse spectral broadening is considered as four-wave mixing (FWM) [3]. In addition, it is worth noting that once WB appears, it occurs synchronously in both time and frequency domain no matter the oscillation of which curve appears ahead. From Figure 2, we can determine the distance where WB happens is at a fiber length about 70m from the valley of curve oscillation.

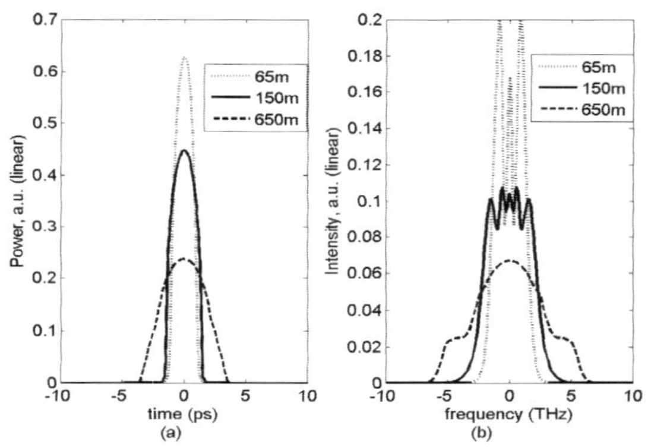

Figure 3. Time and spectral profiles of the pulse at three different distances during the evolution.

In Figure 3, the pulse temporal and spectral profiles at three typical distances (65m, 150m, and 650m) are chosen from the evolution for comparison. We see pulse broadens with the distance in both time and frequency domain. However, the profiles of the pulse spectral intensity vary much more severely than that of the pulse temporal intensity, which reveals the dynamics of self-similar pulse evolution more clearly. As illustrated, the process of the pulse spectral evolution can be classified into about three stages: In the initial stage (from 0m to 65m), it is dominated by the expansion of ripples in the central part of pulse shown at 65m, which is the unique manners of the pulse spectral broadening caused by self-phase modulation (SPM); In the last stage (after 650m), it is dominated by the widening of sidelobes in the pulse wings shown at 650m, which is the particular manners of the pulse spectral broadening caused by FWM; in the middle stage (from 65m to 650m), it is dominated by both the expansion of ripples in the central part of pulse and the widening of sidelobes in the pulse wings shown at 150m, which signifies that both SPM and FWM take effect.

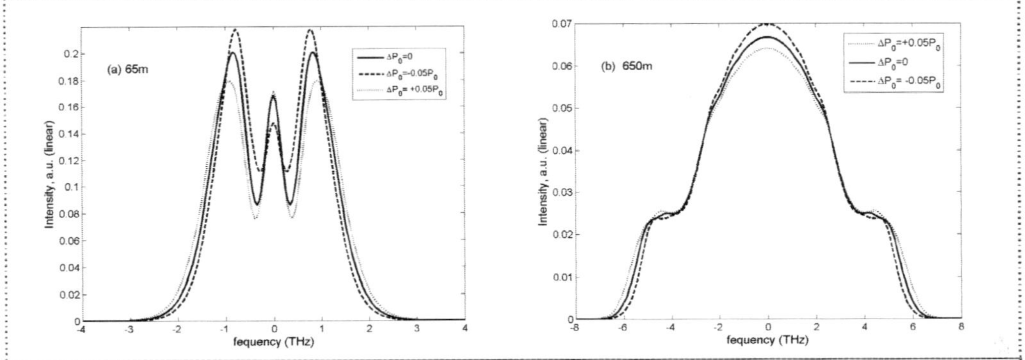

Figure 4. Vibrating spectral profiles of the pulse at 65m (a) and 650m (b) with amplitude noise added

The results can be further verified by adding amplitude noise to the seed pulse as shown in figure 4. The amplitude jitter boundary is set within 5 percent of the pulse peak power as usual. The response of vibrating spectral profiles to the noise manifests the characteristic spectral structures that just determined by the distinctive spectral broadening mechanisms as discussed above.

4. Conclusion

In conclusion, we have introduced a dimensionless pulse shape function in the frequency domain as well as in the time domain to describe the self-similar pulse evolution. WB will take place for a given picosecond pulse under highly nonlinear condition during the evolution. The pulse spectral profile varies along the propagation reveals that the dynamics of self-similar pulse evolution is governed by the different spectrum broadening mechanisms in the different stages. This is confirmed by the typical pulse spectral structures with amplitude noise added.

References

[1] Barenblatt G I 1996 *Scaling, Self-similarity and Intermediate Asymptotics: Dimensional Analysis* (Cambridge: Cambridge University Press)
[2] Dudley J M, et al. 2007 Self-similarity in ultrafast nonlinear optics *Nat. Phys.* **3** 597-603
[3] Agrawal G 2007 *Nonlinear Fiber Optics* (Berlin: Springer)
[4] Akasaki Hirooka T and Nakazawa M 2004 Parabolic pulse generation by use of a dispersion-decreasing fiber with normal group-velocity dispersion *Opt. Lett.* **29** 498-500
[5] Feng L, et al. 2014 Highly coherent supercontinuum generation with picosecond pulses by using self-similar compression *Opt. Express* **22** 27339-27354
[6] Yuan L, et al. 2012 Wave-breaking-extended fiber supercontinuum generation for high compression ratio transform-limited pulse compression *Opt. Lett.* **37** 2172-2174
[7] Burgoyne B, et al. 2007 Nonlinear pulse propagation in optical fibers using second order moments *Opt. Express* **22** 10075–10090

Nontrivial Characteristics Embedded in Microarray Coexpression Profiles

Chuanyang Yin[1], Huijie Yang[2], Guimei Zhu[3], Bing-hong Wang[4,*]

[1] College of Information and Control, Nanjing University of Information Science & Technology, Nanjing 210044, China
[2] School of Management, University of Shanghai for Science and Technology, Shanghai 200093, China
[3] Department of Physics and Centre for Computational Science and Engineering, National University of Singapore, Singapore 117542
[4,*] Department of Modern Physics and Centre for Nonlinear Science, University of Science & Technology of China, Hefei 230026, China

E-mail: bhwang@ustc.edu.cn

Abstract. We propose the biological robustness as a key criterion in inferring functional networks from microarray coexpression profiles. The networks are constructed by filtering out the small values in the coexpression correlation matrices. The robustness refers to there exists a wide region of the filtering criterion value, in which the constructed networks have almost the same properties, such as the degree distribution, the motif pattern, the edge property and so on. This kind of networks are considered biologically meaningful.

1. Introduction

Understanding gene regulatory mechanism at genome-wide level is one of the essential topics in systems biology [1, 2]. Microarray technology can measure simultaneously the expressions of tens of thousands of genes, which makes it possible to unveil gene expression networks at large scale. However, there exist two challenges in inferring of the networks, namely, the noises and the dimensionality. The noises come from the stochasticity of the gene expression events and the substantial noises incurred by technologies. Generally, we can only conduct limited number of replications in typical microarray experiments. The dimensionality refers to the number of available data points is much smaller than the number of genes. The key to the problems is to introduce restrictions from biological information[5, 52].

Various methods have been proposed to infer the underlying networks [3, 5, 6], such as the genetic algorithms [4], neural networks [7], Bayesian models [8–10], Boolean dynamical models [11–13] and ordinary differential equations-based reverse-engineering algorithms [14–19]. But these methods can not overcome the data shortage, i.e. the dimensionality problem, and/or the computational inefficiency. A predominant and fruitful approach is to study the correlations between gene expression experiments [20–25]. The basic assumptions for the method are that the genes performing associated biological functions should exhibit similar expression patterns and the interactions between genes are linear. This approach allows a relatively scale-free estimate of the gene expression similarities and the identification of positive and negative correlations between genes. However, there are still some debates on the assumptions. And the range-restriction-effect confines the analysis to high-variance-variables, which leads to the indeterminacy of the measure of relatively small fluctuations of gene expressions. Overall, the mentioned methods have to be supplemented by extra restrictions, based upon biological knowledge, to reconstruct correctly the functional networks.

Besides the biological information such as phenotypes, functional classification, and cellular responses, the other resource of biological knowledge is to find the nontrivial characteristics embedded in the correlations between gene expressions. The nontrivial characteristics shed light on the regulatory mechanisms obeyed in the biological processes.

The relations between different elements of a complex system can be described with complex networks. The nodes and edges represent the elements and the relations, respectively. Application of complex network theory to the analysis of biological data has provided insight into the topology of biological networks [26–28].

In this paper, with the concepts in network theory we try to find biologically meaningful characteristics embedded in the correlations of gene expressions, which may provide more clues and restrictions to the reconstruction of the functional networks.

2. Materials and Methods

2.1. Materials

We study two microarray expression profiles of Yeast. One is the microarray expression data of 287 Yeast mutants [29]. The genes expressed in most mutants are considered, resulting a total of 6209 genes. The other is the response expressions of Yeast cells to environment temperature changes [30]. Totally 6090 genes expressed in most of the experiments are selected.

2.2. Gene Expression Correlations

The Pearson correlation coefficient is widely used as a metric to calculate the similarity estimations between different genes. We denote the expression measurements with X_{nm}, where $m = 1, 2, \cdots, M$ are the responses to the totally M different conditions, $n = 1, 2, \cdots, N$ the labels of the genes. $X_n = \{X_{n1}, X_{n2}, \cdots, X_{nM}\}$ is called the expression series of the gene i. The coexpression correlation matrix can be constructed as,

$$R_{ij} = \frac{\sum_{m=1}^{M} \left(X_{im} - \bar{X}_i\right)\left(X_{jm} - \bar{X}_j\right)}{\sqrt{\sum_{m=1}^{M} \left(X_{im} - \bar{X}_i\right)^2} \cdot \sqrt{\sum_{m=1}^{M} \left(X_{jm} - \bar{X}_j\right)^2}}, \tag{1}$$

where $\bar{X}_i$ and $\bar{X}_j$ are the averages of the expressions for genes i and j, respectively. To filter out the positive failures due to noises in R, we can introduce a threthold r_c. It is widely accepted that the correlation values larger than r_c correspond to edges with high confidence. In this way, we can construct the functional network A, whose adjacency matrix reads,

$$A_{ij} = \begin{cases} 0 & i = j, \\ 1 & R_{ij} \geq r_c, \\ -1 & R_{ij} \leq -r_c, \\ 0 & \text{otherwise}. \end{cases} \tag{2}$$

How to determine the criterion r_c is the key task in the procedure. In literature [31, 32], a special value is assigned to r_c, at which the degree distribution of the constructed network obeys a power-law and the number of edges is as large as possible. In Ref. [33], it is found that with the increase of the filtering value there exists a critical point at which the distribution function of the nearest neighbor level spacings for the spectrum of the constructed network changes from Wigner to Poisson distribution (the two extremes $\beta = 1$ and $\beta = 0$ for the distribution function, $\sim s^{\beta} \cdot e^{-s^{\beta+1}}$). This critical point is proposed as the criterion r_c. Carter *et al* propose a P-value criterion to construct the networks [34]. Independently permuting the components of each gene expression series and recalculating all the correlations, the resulting correlation distribution can be used to estimate a P-value for each correlation coefficient in R. The P-value threshold is 10^{-5}. While in a recent study [35], a maximum value of r_c is used at which the network keeps connected.

This kind of characteristic transition-based strategies can help us to find useful information on the regulatory mechanisms, but different criteria may lead to different yet artificial results and conclusions. Actually, we have not enough evidences to support the idea that the gene regulatory networks should exist at the transition states. In this paper, we consider the evolutions of the properties of networks constructed with different values of r_c. We examine systematically the degree distribution, the survival edges and the motif patterns of the constructed networks. The nontrivial characteristics in the evolutions may provide more biologically meaningful restrictions, which may lead to a much more reasonable criterion r_c.

2.3. Robustness

Besides the characteristic transition-based criterion, the robustness may be a much more reasonable benchmark for constructing networks. Regarding the correlation coefficient R_{ij} as the essentiality of the corresponding edge between genes i and j. It is reasonable to assume that the real networks, from high essentiality core as a seed to the final real functional network, obey the same set of biologically meaningful rules. The fractal structures of many kinds of biological networks support this assumption [36–38]. Namely, a constructed network reflecting correctly the regulatory relations should be criterion-independent. The impacts of the criterion value, i.e., the artificial characteristics, should be neglectable.

The value of r_c determines the characteristics of the constructed functional network. If r_c is extremely small, the biological meaningful patterns will be submerged by the artificial edges from weak correlations. With the increase of r_c, the artificial edges are filtered out step by step and the biological meaningful patterns tend to dominate the network structures. However, an unreasonable large value of r_c will destroy the patterns by eliminating most of the meaningful edges. Hence we should expect a wide region of r_c in which the constructed networks can capture the characteristics of the real functional networks. If the networks in this region have almost same characteristics, we can conclude that the constructed networks are governed by the same law, i.e., criterion-independent. This range of r_c indicates robustness of organism.

The degree distribution function (DDF) of the constructed networks is used as the measure to detect the criterion-independent characteristics. With the increase of the criterion r_c the width of degree distribution, denoted with K, will decrease monotonically. The criterion-independent implies that the form of DDF keeps unchanged. This is a kind of scale-invariance (the scale is the width K). The DDF can then be expressed as [39–41],

$$P(k) \sim \frac{1}{K^\sigma} \cdot F\left(\frac{k}{K^\sigma}\right), \tag{3}$$

where k is the degree and σ the self-similarity exponent. For a power-law distribution, it equals to the power-law exponent, while for a Gaussian distribution the self-similarity property is trivial, i.e., $\sigma = 0.5$.

2.4. Survival Edges

With the increase of the criterion r_c more and more edges with low confidence are discarded and the survival edges are supposed to be biologically meaningful. Hence, from this evolution process we can separate the structure patterns in the real functional network from the artificial edges. And then we can find the statistical characteristics of the artificial edges and the structure patterns of the real network.

For the edge $A_{ij} = 1$, we can describe its property by using the properties of nodes i and j at the two ends of the edge. Herein, we introduce four quantities,

$$\begin{array}{c} k_{i-j} = |k_i - k_j|\,, k_{i+j} = k_i + k_j; \\ C_{i-j} = |C_i - C_j|\,, C_{i+j} = C_i + C_j, \end{array} \tag{4}$$

where k_i, k_j and C_i, C_j are the degrees and the clustering coefficients of the nodes i and j, respectively. $k_{i-j}, k_{i+j}, C_{i-j}$ and C_{i+j} are called edge degree difference (EDD), edge degree summation (EDS), edge clustering difference (ECD) and edge clustering summation (ECS), respectively.

2.5. Motifs

Motifs are some special subgraphs containing several connected nodes [42]. These subgraphs occur with significant high probabilities compared with that in the corresponding randomized networks. This concept is firstly introduced by Milo *et al* to describe the local structures of complex networks [43]. Detailed works show that the gene transcription, protein-protein interaction and metabolic networks can be characterized with different kinds of motifs. And the motifs can have different dynamical characteristics, which in turn can act as bricks in biological processes. In this paper we consider the motifs in the constructed networks with different values of r_c. The changes of the motif patterns imply the changes of biological functions for the networks.

We consider all the subgraphs containing three nodes. For each constructed network, we detect all the possible three-node subgraphs in the networks and reckon the numbers of the subgraph patterns, respectively. Then generating an ensemble of complete random networks with same numbers of edges and nodes, from which we can obtain the average numbers of the subgraphs. In order to find occurring significance of a certain subgraph, a statistical quantity Z-score is often used, which reads,

$$Z_i = (N_i^{real} - \left\langle N_i^{rand} \right\rangle) \,/\, std(N_i^{rand}), \tag{5}$$

where N_i^{real} is the occurring number of subgraph i in the real network, $\left\langle N_i^{rand} \right\rangle$ and $std(N_i^{rand})$ the average and the standard deviation of the occurring numbers of subgraph i in the ensemble of randomized networks, respectively. If the Z-score of a subgraph is high, we call the subgraph a motif of the network. For convenience, Z-score is normalized to z_i as [44],

$$z_i = \frac{Z_i}{\sqrt{\left[\sum_j Z_j^2\right]}}. \tag{6}$$

The summation covers Z-scores of all the possible subgraphs.

3. Results

At different values of the criterion r_c we construct the networks for the cells response and the mutants profiles, respectively. Fig.1 presents several typical DDFs for the cells response networks at different r_c. With the increase of r_c we can find a wide region of r_c in which the DDF obeys power-law. The region is $r_c \in (0.42, 0.92)$. For the mutants networks, we have similar results as shown in Fig.2. The region where the DDF obeys power-law is a slightly different as, $r_c \in (0.47, 0.92)$. Out of the regions the DDF will depart significantly from power-law behavior.

Figure 3 presents the slopes of the DDFs for networks at different r_c. We can find that though the DDFs have power-law behaviors, at the sub-regions near the upper and lower borders the slope changes significantly from 0.55 to 1.25. For the cells response networks there exists a

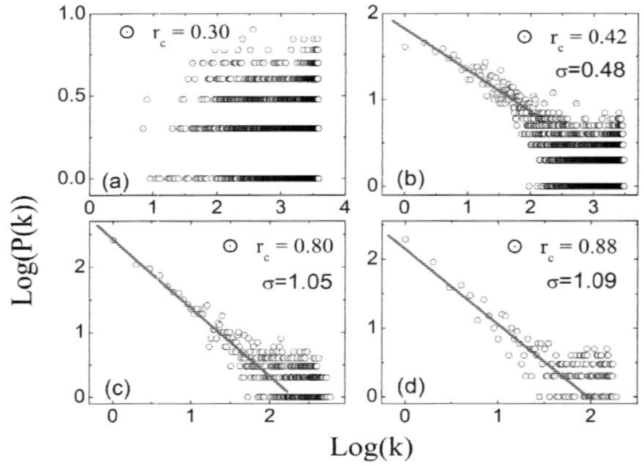

Figure 1. (Color online) Typical DDFs for cells response networks constructed with different values of r_c. In a wide region of $r_c \in [0.42, 0.92]$ the DDFs have power-law behaviors.

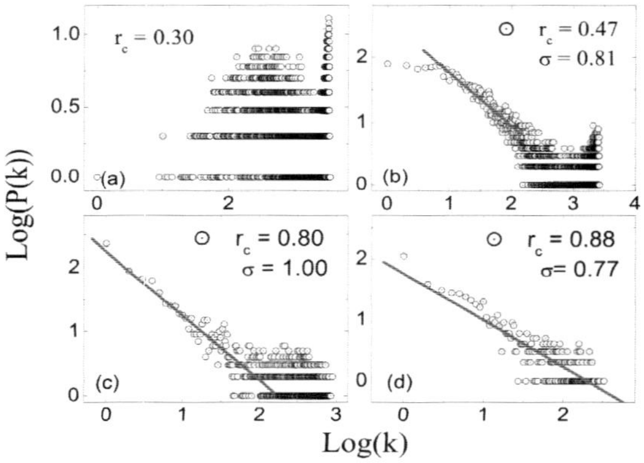

Figure 2. (Color online) Typical DDFs for mutants networks constructed with different values of r_c. In a wide region of $r_c \in [0.47, 0.92]$ the DDFs have power-law behaviors.

region, $r_c \in [0.63, 0.88]$, in which the slope keeps almost unchanged, $\sigma = 1.00 \pm 0.05$, i.e., criterion-independent. The region for mutants networks is, $r_c \in [0.62, 0.82]$. That is to say, the networks in the regions have self-similarity characteristics. These robust networks should capture the real functional patterns.

With the increase of r_c, the artificial edges are filtered out step by step. Using the quantities of EDD, EDS, ECD and ECS we try to find the characteristics of the biological meaningful edges (survival edges) and the artificial edges. The simulations, which start with the same network constructed by the Pearson correlation coefficient, are performed in two methods, i.e., our r_c based filtering process and randomly discarding edge process. At the beginning of the simulations, we build an initial network with a very small r_c, say, r_c=0.05. The filtering processes are performed with the ascending values of r_c, so we can get a series of networks and

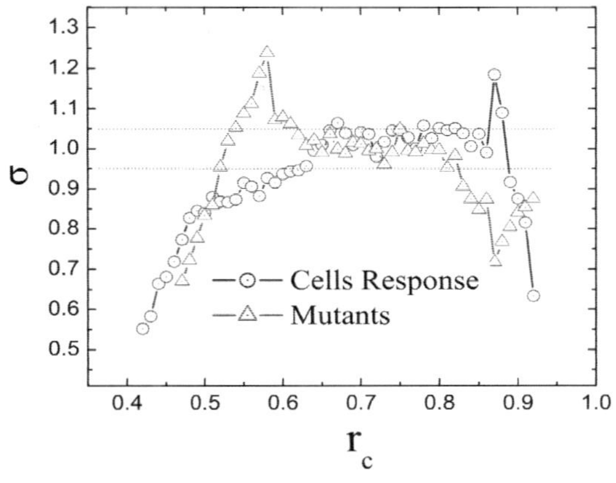

Figure 3. (Color online) The relation between the slope σ and the criterion r_c. Near the borders σ changes significantly from 0.55 to 1.25. We can find an region in which σ keeps almost unchanged as 1.00 ± 0.05. The regions for the cells response and mutants networks are, $[0.63, 0.88]$ and $[0.62, 0.82]$, respectively.

the number of abandoned edges in the previous step, respectively. Meanwhile, the randomly filtering processes followed as a reference, which discard the same number of randomly selected edges. At each of the step, we calculate EDD,EDS,ECD,and ECS for each removed edge and calculate the distribution function of the results.

Fig. 4 shows several typical results for cells response networks. It is found that at a small value of r_c , the filtering process is significantly different from the randomly discarding process. The edges with large values of EDD and ECD prefer to be filtered out firstly compared with the corresponding random process. That is, an edge between two nodes with close properties (degree and clustering coefficient) tends to survive, which may indicate these edges have high confidence of biological meaning. We can find also that edges with large EDS tend to be conserved. We can not find distinctive differences between the ECS distributions for the reconstruction procedure and the random one.

When the value of r_c becomes large enough, e.g., $r_c = 0.8$ and 0.9, the filtering process tends to become indistinguishable from the random one. At this stage, the biologically meaningful edges are discarded, which will destroy the real network patterns.

Similar conclusions can be reached for the mutants networks (not shown).

There are totally 7 patterns for the subgraphs containing three nodes, as shown in Fig.5a. The red dashed and black solid edges denote the positive and negative relations between two genes. Fig. 5b and Fig. 5c exhibit the Z-scores, z_i, for the subgraphs numbered ID-1 to ID-7 in the cells response and mutants networks, respectively.

Generally, the subgraphs numbered ID-4 and ID-6 are motifs in the wide robust region of r_c. When r_c reaches the upper border of the robust region, graph ID-6 are not motif for the cells response networks. And even the original motifs are replaced by the graphs ID-1,ID-2 and ID-3 as new motifs in the mutants networks. This characteristic not only verifies the existence of a wide robust region of r_c in which the constructed networks are biologically meaningful, but also provides a criterion to determine the upper border of r_c.

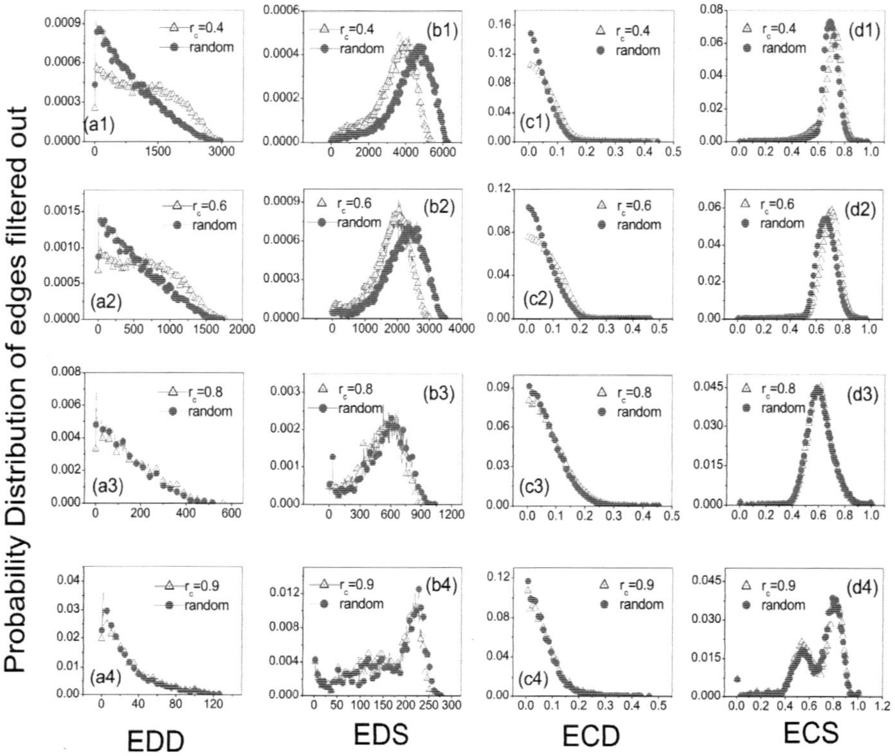

Figure 4. (Color online) Properties of the filtered out edges with the increase of r_c for cells response networks. (a1)-(a4) the edges with large EDD values tend to be filtered out firstly. The edges between nodes with close degrees are much more biologically meaningful; (b1)-(b4) the edges with small values of EDS tend to be filtered out firstly. The edges between nodes with large degrees can survive; (c1)-(c4) the edges with large ECD values tend to be filtered out firstly. The edges between nodes with close clustering coefficients may have high confidence of biological meaning; (d1)-(d4) there is no significant difference between the distributions of ECS for the random and the reconstruction procedures. Near the upper border of the robust region, e.g., $r_c = 0.8$ and 0.9, the nontrivial characteristics in the reconstruction procedure become undistinguishable from the random one.

4. CONCLUSION

The reconstruction of regulatory networks from microarray data is one of the essential problems in systems biology. To get useful information, we must overcome two challenges, the noises and the dimensionality. Various methods have been proposed, but no method can reach valuable results without the supplementary of biological knowledge. Besides the biological information from experiments, finding nontrivial characteristics embedded in microarray data can also provide much more useful information on regulatory mechanisms.

In this paper, the measures of complex networks such as the degree, edge properties and

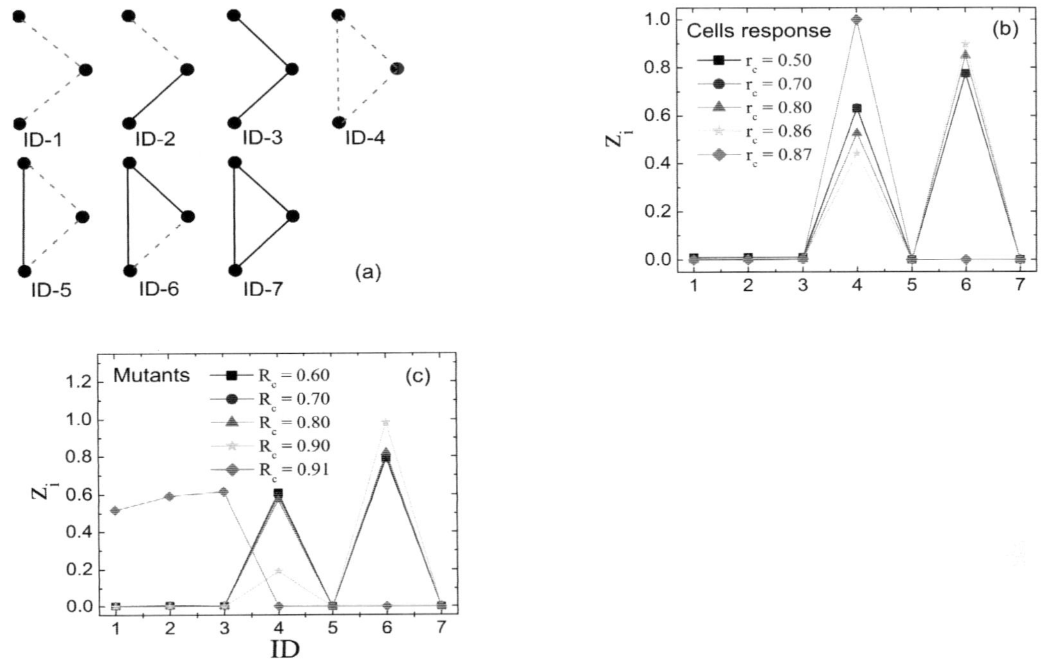

Figure 5. (Color online) (a) All the possible types of subgraphs containing three connected nodes. The red dashed and black solid edges denote up-regulatory and down-regulatory relations respectively; (b) Z-scores for the total 7 subgraphs in the cells response networks. Near the upper border of the robust region only the subgraph ID-4 survives as motif; (c) Z-scores for the total 7 subgraphs in the mutants networks. Near the upper border of the robust region the subgraphs ID-4 and ID-6 are replaced by the subgraphs ID-1, ID-2 and ID-3 as motifs.

motif patterns are used to find the biologically meaningful characteristics embedded in the gene coexpression correlation matrices. These measures provide characteristics of the constructed networks from microscopic to macroscopic scales.

We find that there exists a wide region of the criterion r_c, in which the constructed networks are robust, namely, the DDF keeps unchanged. This kind of scale-invariance is criterion-independent and biologically meaningful.

In the robust region, with the increase of r_c the filtered out edges are significantly different from the ones randomly selected. This nontrivial filtering procedure implies its biological meaning. When r_c reaches the upper border of the robust region, the significance of this procedure tends to be indistinguishable from the random one.

The existence of the robust region is confirmed by the motif patterns in the constructed networks. In the main part of the region, the motifs keep unchanged, i.e, the subgraphs ID-4 and ID-6 are motifs. Near the upper border, ID-6 will not be a motif for cells response networks and even the graphs ID-1,ID-2, ID-3 become motifs instead of the original motif patterns ID-4 and ID-6 for the mutants networks.

To conclude, based upon the biological robustness of organism, the determination of criterion r_c is a natural process and the robust region r_c should lead much more reliable networks.

our method may provide a new perspective in referring functional networks from microarray coexpression data.

Acknowledgments
This work is partially supported by the National Natural Science Foundation of China (Grant No. 61273229), the Research fund of Nanjing University of Information Science & Technology (20110395). The authors are grateful to B. W. Li for useful discussions.

References

[1] Cho K, Choo S, Jung S, Kim J, Choi H and Kim J 2007 *IET Syst. Biol.* **1** 149
[2] Sprinzak D and Elowitz M 2005 *Nature* **438** 443
[3] Bansal M, Belcastro V, Ambesi-Impiombato A and Bernardo D 2007 *Mol. Syst. Biol* **3** 78
[4] Wahde M and Hertz J 2000 *BioSystems* **55** 129
[5] Hecker M, Lambecka A, Toepferb S, Somerenc E and Guthkea R 2009 *BioSystems* **96** 86C103
[6] Werhli A, Husmeier D 2008 *Journal of Bioinformatics and Computational Biology* **6** 543-72
[7] D'Haeseleer P, Liang S and Somogyi R 2000 *Bioinformatics* **16** 707
[8] Hartemink A, Gifford D, Jaakkola T and Young R 2001 *in Pac. Symp. Biocomputing* **6** 422
[9] Yu J, Smith V, Wang P, Hartemink A and Jarvis D 2004 *Bioinformatics* **20** 35949
[10] Dojer N, Gambin A, Mizera A, Wilczynski B and Tiuryn J 2006 *BMC Bioinformatics* **7** 249
[11] Akutsu T, Miyano S and Kuhara S 1999 *Pac. Symp. Biocomput.* **4** 17
[12] Akutsu T, Miyano S and Kuhara S 2000 *Bioinformatics* **16** 727
[13] Martin S, Zhang Z, Martino A. and Faulon J 2007 *Bioinformatics* **23** 866
[14] Someren E, Wessels L and Reinders M 2000 *Proc. Int. Conf. Intell. Syst. Mol. Biol.* **8** 355
[15] Yeung M, Tegne J and Collins J 2002 *Proc. Natl. Acad. Sci. USA* **99** 6163
[16] Gardner T, Bernardo D, Lorenz D and Collins J 2003 *Science* **301** 102
[17] Someren E, Vaes B, Steegenga W, Sijbers A, Dechering K and Reinders M 2006 *Bioinformatics* **22** 477
[18] Bonneau R,Reiss D,Shannon P, Facciotti M, Hood L, Baliga N and Thorsson V 2006 *Genome. Biol.* **7** R36
[19] Bansal M and Bernardo D 2007 *IET Syst. Biol.* **1** 306
[20] Elkon R, Linhart C, Sharan R, Shamir R and Shiloh Y 2003 *Genome Res.* **13** 773
[21] Crescenzi M and Giuliani A 2001 *FEBS Lett.* **507** 114
[22] Basso K, Margolin A, Stolovitzky G, Klein U, Dalla-Favera R and Califano A 2005 *Nat. Genet.* **37** 382
[23] Butte A and Kohane I 2000 *Pac. Symp. Biocomput.* **5** 418
[24] Faith J, Hayete B, Thaden J,Mogno I, Wierzbowski J, Cottarel G, Kasif S, Collins J and Gardner T 2007 *PLoS Biol.* **5** e8
[25] Davidson E, McClay D and Hood L 2003 *Proc. Natl. Acad. Sci. USA* **100** 1475
[26] Watts D 1999*Small Worlds* (Princeton: Princeton University Press)
[27] Albert R and Barabasi A 2002 *Rev. Mod. Phys.* **74** 47
[28] Dorogovtsev S and Mendes J 2003 *Evolution of Networks* (New York: Oxford University Press)
[29] Hughes T et al. 2000 *Cell* **102** 109
[30] Gasch A et al. 2000 *Mol. Biol. Cell* **11** 4241
[31] Zhang B and Horvath S 2005 *Stat. Appl. Gene. Mol. Bio.* **4** 17
[32] Rhoa K, Jeongb H and Kahng B 2006 *Physica* A **364** 557
[33] Luo F,Zhong J, Yang Y and Zhou J 2006 *Phys. Rev.* E **73** 031924
[34] Carter S, Brechbuhler C, Griffin M and Bond A 2004 *Bioinformatics* **20** 2242
[35] Sun L, Li M, Jiang L and Tan L 2007 *Physica* A **384** 739
[36] Song C, Havlin S and Makse H 2005 *Nature* **433** 6392
[37] Song C, Havlin S and Makse H 2006 *Nature Physics* **2** 275
[38] Yang H, Yin C, Zhu G, Li B 2008 *Phys. Rev.* E **77** 045101
[39] Scafetta N and Grigolini P 2002 *Phys. Rev.* E **66** 036130
[40] Yang H, Zhao F, Zhang W and Li Z 2005 *Physica* A **347** 704
[41] Zhao F, Yang H and Wang B 2007 *J. Theor. Bio.* **247** 645
[42] Alon U 2006 *An Introduction To Systems Biology:Design Principles Of Biological Circuits* (Chapman and Hall/CRC)
[43] Milo R, Shen-Orr S, Itzkovitz S, Kashtan N, Chklovskii D and Alon U 2002 *Science* **298** 824
[44] Eom Y, Lee S and Jeong H 2006 *J. of Theor. Biology* **241** 823
[45] Yang H, Zhao F, Qi L and Hu B 2004 *Phys. Rev.* E **69** 066104
[46] Zhao F, Yang H and Wang B 2005 *Phys. Rev.* E **72** 046119

[47] Yang H, Zhao F and Wang B 2006 *Physica* A **364** 544
[48] Yang H, Zhao F and Wang B 2006 *Chaos* **16** 043112
[49] McGraw N and Menzinger M 2007 *Phys. Rev.* E **75** 027104
[50] Bandyopadhyay J and Jalan S 2007 *Phys. Rev.* E **76** 026109
[51] Ivanov P, Amaral L, Goldberger A, Havlin S, Rosenblum M,Struzik Z and Stanley H 1999 *Nature* **399** 461
[52] Wang Y, Wu C, Ji Z, Wang B, Liang Y 2011 *PLoS ONE* **6**(5) e20060

Power-law spectrum and small-world structure emerge from coupled evolution of neuronal activity and synaptic dynamics

Hong-Li Zeng[1], Chen-Ping Zhu[1,*], Yan-Dong Guo[1,2], Ao Teng[1], Jing Jia[1], Hui Kong[1], Rui Zhou[1], Juan-Ping Yang[1] and Su-Quan Li[1]

[1] College of Science,Nanjing University of Aeronautics and Astronautics,Nanjing 210016, China

[2] College of Electronic Science and Engineering, Nanjing University of Posts and Telecommunications, Nanjing 210046, China

E-mail: oldpigman1234@126.com

Abstract. A co-evolutionary neuronal network model based on previous ones is proposed, and both functional and structural properties are numerically calculated. Recent experiments have revealed power-law behavior in electrocorticogram (ECoG) spectrum related with synaptic plasticity and reorganization. In the present neuronal network model, the network starts its evolution from the initial configuration of random network which is the least biased and without special structure, and the interaction rules among neurons are modified from both models by Bornholdt's and Arcangelis' groups to simulate the process of synaptic development and maturation. The system exhibits dynamic small-world structure which is the result of evolution instead of the assumption beforehand. Meanwhile, the power spectrum of electrical signals reproduces the power-law behavior with the exponent 2.0 just as what is experimentally measured in ECoG spectrum. Moreover, the power spectrum of the average degree per neuron over time also exhibits power-law behavior, with the exponent 2.0 again over more than 5 orders of magnitude. Different from previous results, our network exhibits assortative degree-degree correlation which is expected to be checked by experiments.

1. Introduction

In living organisms, information processing is generally performed by large networks of interacting neurons with huge complexity[1]. However, the level of the complexity exceeds what can be coded by the observed number of genes. So, evolving principles is quite important in network construction. Self-organization and adaptation processes are thought to exist throughout the lifetime of a network, with learning as its major function. It has been reported that, self-organized criticality (SOC) universally exists in natural phenomena [2–9]including neural networks [1, 10–24]. In most studies, the neural network in the brain is supposed to be a small-world (SW) one [25, 26]. Many neural network models have been applied to or based on it. In this paper, we argue that SOC and SW should emerge simultaneously in the same process since no evidence shows that their appearance should be one after another. And we propose a co-evolutionary model to support this opinion.

Based on self-organization of degree of connectivity, Bornholdt et al. [27] have proposed a neural network model. In their model, the network topology changes slowly according to a local rewiring rule motivated by Hebbian [28], activity-dependent synaptic development, i.e., neighbor neurons whose activities are correlated would develop a new connection, and uncorrelated neighbors tend to disconnect. It is a co-evolutionary process, as the node states and their correlations update simultaneously. As a result, robust self-organization of the network towards the order-disorder transition occurs, which is independent of initial conditions, robust against thermal noises, and dose not require fine tuning of parameters.

Recently, by taking into account the brain plasticity, Arcangelis, Perrone-Capano, and Herrmann (APH) [29] proposed another neural network model. This model is based on an electrical network on a square lattice with the features of threshold firing and activity-dependent synapse strengths. The system exhibits an avalanche activity in a power-law distribution. The power spectrum of the electrical signal could reproduce the $1/f$ power-law behavior observed in experimental EEG spectra.

These two models, Bornholdt and APH ones, are distinct from each other, but they both exhibit the feature of SOC. And they are both consistent with Bak's [30] guess that the brain should be critical, not subcritical or supercritical. Actually, SOC is an important discovery in statistical physics[30], and has been found in many complex systems. For instance, earthquakes, forest fires and etc.. Under general conditions, these dynamical systems, consisting of many interacting elements, can organize themselves into a critical state in which power-law distribution of avalanche sizes and $1/f$ noise in power spectrum in time-series appear simulaneously, and the dynamical response of the system to external environment gets its maximal level in such non-equilibrium steady states. In particular, these features emerge without any significant tuning of the system from the outside, that is to say, variable parameters in the model could be changed widely without affecting the emergence of critical behavior.

All the typical features of SOC have been observed in the Bornholdt and APH models. Like many other models, the initial topologies of their neural networks are supposed to be biased ones, e.g., square lattice or SW. However, there is no evidence so far supporting their assumptions. Here, we ague that, neural network should evolve from a random network, which has less artificial and biased rules. Structural and functional features, observed in experiment, should emerge simultaneously during the same single co-evolutionary process.

In this paper, we propose a co-evolutionary model starting its evolution from a random network, developed from Bornholdt and APH models. By analyzing its structural properties , we found that, two power-law behaviors emerge in the co-evolutionary process, i.e., power spectra of the electrical signals and average degree per node over time. In the following, we will first introduce some variables and detailed mechanisms appeared in Bornholdt and APH models. Then, we propose our modified model. At last, numerical results will be discussed and compared with experimental ones.

2. Model

2.1. Bornholdt model

In Bornholdt model, links between nodes is represented by W_{ij}, which is randomly drawn from a uniform distribution $W_{ij} \in [-1, 1]$. The couplings between neuron is and js are asymmetric, i.e., $W_{ij} \neq W_{ji}$. The asymmetry has its fundamental significance. First, the transfer of signals though synapse is unidirectional. Second, the neural system must be robust against noises and meanwhile must be sensitive to inputs [31]. In previous studies, a winner-less competition (WLC) principle [32] was proposed. According to this principle, the non-symmetric links could be used to deal with the contradiction of signals and noises. The key point is that, signals and noises have different features, and they influence the dynamics of the system in different ways [31].

For each neuron in the network, it has its own state $\sigma_i = \pm 1$, which will update according to a stochastic Little dynamics [27] based on inputs received from its neighbor neurons in last time step:

$$\begin{aligned} \Pr ob[\sigma_i(t+1) = +1] &= g_\beta(f_i(t)) \\ \Pr ob[\sigma_i(t+1) = -1] &= 1 - g_\beta(f_i(t)), \end{aligned} \tag{1}$$

$$g_\beta(f_i(t)) = \frac{1}{1 + e^{-2\beta f_i(t)}}, \tag{2}$$

$$f_i(t) = \sum_{j=1, j\neq ij}^{N} W_{ij}\sigma_j(t) + \theta_i, \tag{3}$$

where β is the inverse temperature and θ_i is the threshold. Here $\theta_i = -0.1 + \gamma$ and γ is a small random noise term from a Gaussian distribution with the width of $\varepsilon = 0.1$ and the mean of $\mu = 0$. These terms are introduced from the slow fluctuations observed in biological neural systems [27, 33].

In Bornholdt model, the quantity $C_{ij}(\tau)$ is defined as the average correlation of a pair (i, j) of neurons over a time interval τ:

$$C_{ij}(\tau) = \frac{1}{\tau + 1} \sum_{t=t_0}^{t_0+\tau} \sigma_i(t)\sigma_j(t). \tag{4}$$

This could be used as an order parameter to guide the topology rewiring. If the activities of two neighboring neurons are on average highly (anti)correlated, $C_{ij}(\tau)$ will be large and the two neurons will set up a link. Otherwise, the link will be removed. Obviously, it is a local rewiring rule. The updates of node states couple with their correlations, and this leads to a co-evolutionary process between node dynamics and network connections. As pointed out by Bak et al. [30], such kind of simple local interactions could spontaneously induce complexity.

2.2. APH model

Arcangelis et al. [29] proposed a neuronal firing model. The neurons are simply imposed on a square lattice. Therefore, there is a neuron on each site and bonds on the lattice acts as couplings between neurons. Each neuron (node) has its own potential V_j at site j, and each synapse (link) has its own conductance g_{ij}. When the potential is above a threshold, i.e., $V_i > V_{\max}$, the neuron fires. It will generate an action potential and distribute its charges to its connected neighbors according to (for the notations, j is the nearest neighbor of i, and k belongs to j who has the potential of $V_j < V_i$)

$$v_j(t+1) = v_j(t) + v_i(t)\frac{I_{ij}(t)}{\sum\limits_k I_{ik}(t)}, \tag{5}$$

with

$$I_{ij} = (V_i - V_j) \cdot g_{ij}. \tag{6}$$

Once a neuron fires, it will go into a refractory period for a time step, and can not receive charges from other firing neurons until getting out of this period. This could ensure that action potential would not reverberate in the network. The electrical activity is represented by the total current flowing in the network. Supposing the potential of neuron i is equal to or large than v_{max} at time t, it fires follows Eq. (5), then the conductance of links with non-zero current is changes as follows:

$$g_{ij}(t+1) = g_{ij}(t) + \delta g_{ij}(t) \tag{7}$$

with $\delta g_{ij}(t) = k\alpha i_{ij}(t)$ (k, unit constant; α, a tunning parameter). Once the avalanche of firing is over, the bond with non-zero conductance is reduced by $\Delta g = \sum_{ij} \delta g_{ij}(t)/N_b$, N_b is the number of bound with $g \neq 0$. This corresponds to a pruning process of links which mimics the brain development.

2.3. Our co-evolutionary model

It is noticeable that the coupling strength W_{ij} in Bornholdt model plays a similar role as the conductance g_{ij} in APH model in the network evolution process. Here, these two models are combined together by replacing g_{ij} in APH model with the absolute value of W_{ij} in Bornholdt one. Based on the combination of the firing mechanism in APH model and the network evolution process from Bornholdt one, we start from a random network.

The random network we start with has a link probability p, i.e., each pair of neurons can be connected with probability p. The average degree of the network, K, is defined as the average number of nonzero degree of each node (include both incoming and outgoing links). This means our initial network has an average degree $K = pN$, where N is the total number of neuron in the system. Then, the whole model is presented as follows,

(1) Start with a random network with link probability p and a random initial system state with the vector $\vec{\sigma}(0) = (\sigma_1(0), ..., \sigma_N(0))$. The neuron potentials are uniformly distributed between $V_0 - 1$ and V_0.

(2) For each neuron i, choose a random threshold $\theta_i = -0.1 + \gamma$, where γ is a small random noise term from a Gaussian distribution of width $\varepsilon = 0.1$ and mean $\mu = 0$.

(3) Starting from the initial state, calculate the new system state applying Eqs. (1-3) using parallel update. Iterate this for τ time steps. Here we set the inverse temperature to be $\beta = 25$ [27].

(4) For each neuron i, randomly choose neuron j from its neighbors and the neighbors' neighbors (the first and second nearest neighbors). Determine the average correlation $C_{ij}(\tau)$ over the τ time steps according to Eq. (4).

(5) If $C_{ij}(\tau)$ is larger than a given threshold α, node i receives a new link W_{ij} from j with a weight chosen randomly from the interval $W_{ij} \in [-1, 1]$. Note that, weights are asymmetric (here we make the absolute value of W_{ij} as the conductance from i to j. Positive and negative represent excitation and inhibition, respectively). If $C_{ij}(\tau) < \alpha$, the weight of the link W_{ij} is set to be zero.

(6) For each neuron i, if $V_j < V_i$ (where j is the neighbor of i) and they are both out of refractory periodthen i can discharge to j according to Eqs. (5-6). After that, i goes into the refractory period and $V_i = 0$.

(7) Go to step 2 and iterate, using the current state of the network as a new initial state.

To summarize, based on Bornholdt and APH model, we developed a co-evolutionary neural network model. As mentioned above, in our model, the small-world topology should be the result of co-evolution, not a given assumption. We adopt the random network as the initial state, since this kind of network has less biased rules and has no deterministic regular pattern. With less biased and artificial factors, the system could exhibit the effect of co-evolution and SOC more naturally. In detail, nodes' states and interactions should evolve coupled with each other. The coupled evolution process should make a dynamically stable structure and functional properties emerge simultaneously. In addition, this process should be self-organized.

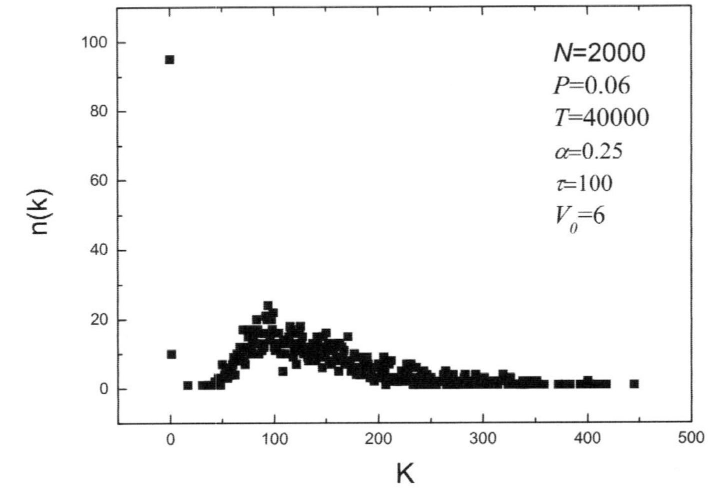

Figure 1. Distribution of degree K for a system with 2000 nodes evolved for $T = 40000$ time steps. The degree of a node is defined as the number of its neighbors, and in this paper it includes both incoming and outgoing links. The initial state is a random network with probability $p = 0.06$. Other parameters: $V_0 = 6$, averaging time window $\tau = 100$, correlation threshold $\alpha = 0.25$.

3. Results and Discussions

3.1. Structural properties produced by our co-evolutionary model

We simulate our model on a random network with $N = 2000$ (number of nodes), $\alpha = 0.25$ (correlation threshold), $\tau = 100$, $V_0 = 6$ and $p = 0.06$. After 40000 time steps, there are 95 neurons being separated from the network. They become isolated nodes ($K = 0$). The distribution of degree, for the final state, is shown in Fig. 1. It is reasonable that a few nodes become isolated, or in other words, dead. From the neurological point of view, during the brain's evolving, there are a number of reasons that could lead to the death of neurons, such as termed apoptosis, environmental toxins, genetic diseases etc. [34–36].

Ignoring those isolated nodes, the distribution of degree, as we can see, is similar to Poisson distribution. One thing should be noted that the degree of a node includes both the incoming and outgoing links, and this rule is also applied to the calculations of clustering coefficient and distribution of $K_{nn}(K)$. Figure 2 presents the clustering coefficient of each node (except the nodes with $K = 0$). The clustering coefficient of a node is defined as the ratio of the number of connections in the neighborhood of this node to the number of connections when the neighbors are fully connected with each other. From this value, we could see whether the neighbor of a node tends to connect with other neighbors of the same node. As shown in Fig. 2, most clustering coefficients range in (0.10, 0.35). Their average is 0.19, which is relatively high for a network. From the distribution of the degree and the clustering coefficient, we can conclude that, the final state is a small-world network.

To understand the network structure deeply, we calculate the distribution of $K_{nn}(K)$ (shown in Figure 3), which is the average degree of the neighbors of the nodes whose degree is equal to K. The whole curve exhibits an upward trend, indicating that the nodes with small degree tend to connect with the nodes also with small degree, and large ones to large ones. So, the network exhibits positive degree-degree correlation. This result can not be proved by the current

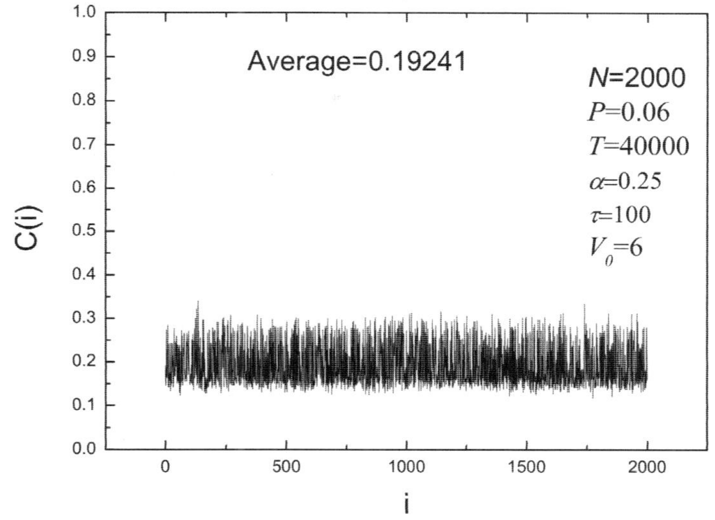

Figure 2. Clustering coefficient $C(i)$ of each node ($i = 1 \sim 2000$, except the nodes with $K_i = 0$). The clustering coefficient of a node is defined as the ratio of the number of connections in the neighborhood of this node to the number of connections when the neighbors are fully connected with each other. The average value of the clustering coefficients is 0.19241.

experiment, but a recent experimental breakthrough shows the possibility, i.e., using brainbow technology, developed by Jean Livet et al [37], one could visualize synaptic circuits by genetically labeling neurons with multiple and distinct colors. In the level of anatomical connections between cortical areas, degree-degree correlation was observed in the latest experimental results [38].

3.2. Functional properties of the network

Figure 4 shows the total current as a function of time, which is the sum of current every time step flowing in the system. At the beginning of evolution, the current is very large, but unstable. Then, it declines rapidly, and after 3300 time steps the current reaches a minimum. After that, it slightly increases, and then oscillates stably. Figure 5 shows the details of the current signal in a short period. Like the ECoG in experiment [39], they vary in the same way. According to the time interval of the experimental ECoG signal [39], every time step in our model stands for 4 ms. If we make it as the sampling period, the corresponding sampling frequency will be 250 Hz. Then we can get the power spectrum (the square of the amplitude of the Fourier transform as a function of frequency) of the current with this sampling frequency (log-log scale for horizontal and vertical coordinates). Surprisingly, it exhibits power-law behavior over more than 5 orders of magnitude. On the left of the figure, the curve shows flat head, and on the right, it show peaks on the tail. These features agree quite well with the experimental results [39]. In particular, the slope (-2.01789) of the curve (Fig. 6), which is the exponent of the function, matches the experimental slopes ranging between -2.27 and -1.54 and with their mean as -1.97[39]. They both exhibit power-law, a feature of SOC. It should be noted that, according to scale-free properties of the power law distribution, the difference of axiss scale does not affect the shape and slope of the curve.

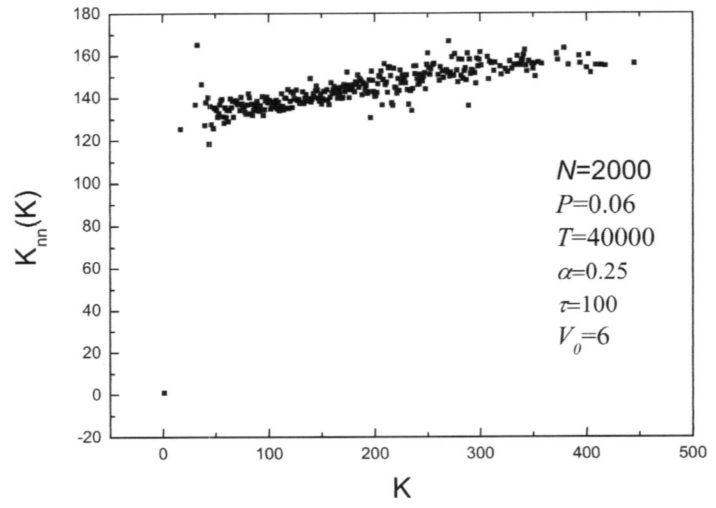

Figure 3. Distribution of $K_{nn}(K)$, which is the average degree of the neighbors of the nodes whose degree is equal to K. An upward trend indicates a positive degree-degree correlation, i.e., the nodes with small (or large) degree tend to connect with the nodes also with small (or large) degree.

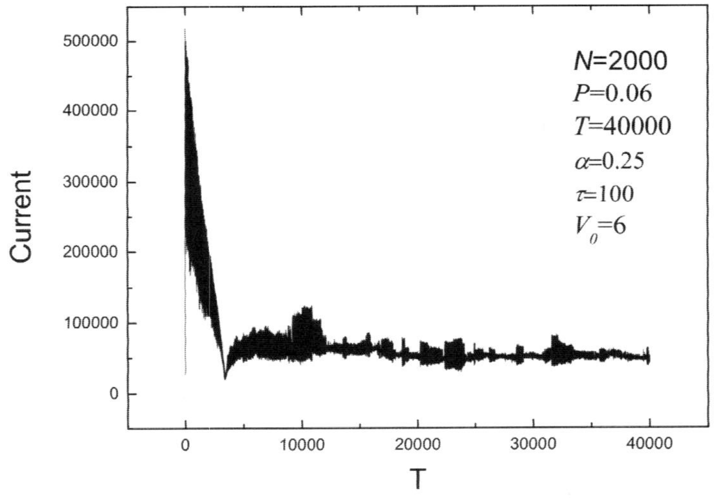

Figure 4. Total current flowing in the system as a function of time, which is the sum of current every time step flowing in the whole system. After about 3300 time steps, it reaches a stable oscillating state.

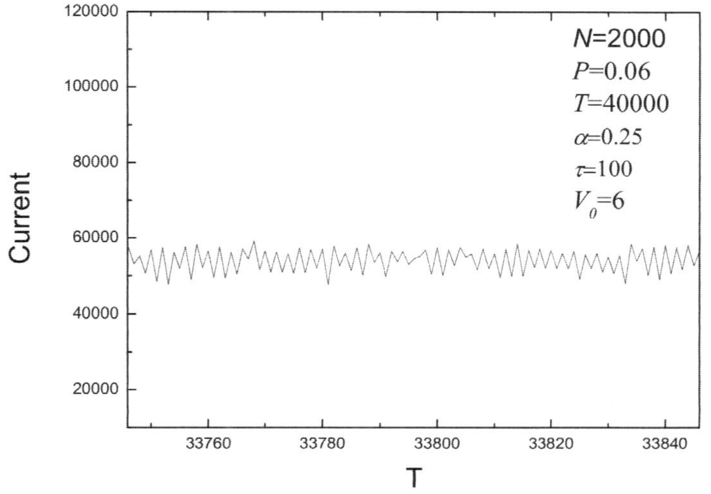

Figure 5. Local details of the total current shown in Fig. 4 for a short period. Comparing with the experimental ECoG signal [39], every time step in our model is estimated to be 4 ms.

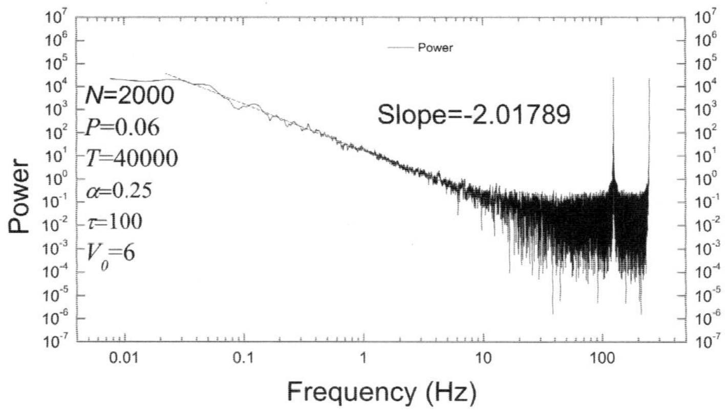

Figure 6. Power spectrum of the simulated total current, which is the square of the amplitude of the Fourier transform as a function of frequency for the current signal. The sampling frequency is 250 Hz, corresponding to 4 ms. Log-log scale is chosen for both horizontal and vertical coordinates. A power-law behavior over more than 5 orders of magnitude emerges. The straight (red) line shows the linear fit for the middle part of the curve.

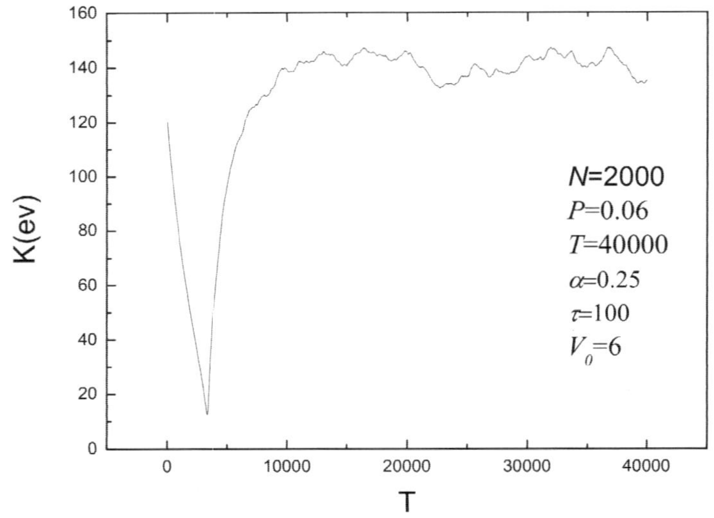

Figure 7. The average degree per neuron changes as a function of time. After about 3200 time steps, it reaches a stable oscillating state, being consistent with the current (Fig. 4).

3.3. Structural properties during co-evolutionary process

Our initial network is a random network consisting of 2000 nodes, and the link probability is 0.06. It is a sparse network. In Bornholdt model [27], the absolute average correlation is studied under different conductivities K. It is found that the correlation is large for networks with small connectivity, and small for networks that are densely connected. This quantity is used as a criterion for rewiring process, and it makes the rewiring rule reach a balance, i.e., links tend to be created at strong correlation, and destroyed at weak correlation. Thus, without global information, the local rewiring rule, synaptic evolving depending on averaged correlation between neurons, could guide the whole network to a convergent state. Figure 7 shows the average degree per neuron changes as a function of time. Obviously, after 3200 time steps, the average degree reached its minimum, being consistent with the current (Fig. 4). This is the time at which 4.8 percent nodes are separated from the network and became isolated nodes. After that, the average degree rises up to about 140 rapidly, and oscillates stably around it. This demonstrates that local rewiring mechanism could control global dynamical properties in a large network [27]. Whether nearby neurons establish a connection between them is only dependent on the degree of correlation between their activities. However, as a result, the entire network could evolve into a stable state. It is a key point of SOC. The final state is not static, and it changes slightly over time. We think it is these slight changes that create a variety of functions of the brain, e.g., learning and memory. To figure out the inherent mechanism imbedded in such changes, we did Fourier transform for the evolution of this average degree, and got its power spectrum (Fig. 8). Interestingly, it exhibits power-law behavior with the exponent 1.95 (close to 2.0) over more than 5 orders of magnitude. And the linear of the middle part of the curve is quite well. Like the current, the average degree changes also show $1/f^2$ noise. This is another evidence and presentation of SOC.

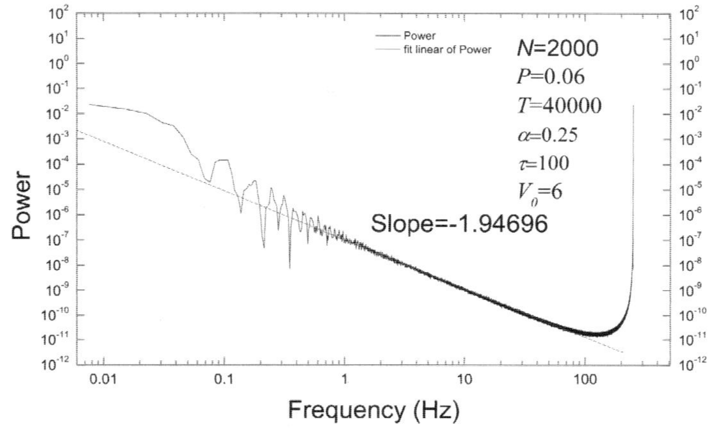

Figure 8. Power spectrum of the average degree per neuron, which is the square of the amplitude of the Fourier transform as a function of frequency for the average degree signal in Fig. 7. The sampling frequency is 250 Hz. Log-log scale is chosen for both horizontal and vertical coordinates. A power-law behavior like that in Fig. 6 is also observed. The straight (red) line shows the linear fit for the middle part of the curve.

3.4. Ranges of parameters

As we mentioned before, for SOC, variable parameters could be changed widely without influencing the emergence of critical behavior. Next we focus on the ranges of parameters in our model. By performing a large number of numerical simulations, we get the following ranges

$$\alpha \sim [0.1, 0.6],$$

$$p \sim [0.06, 1],$$

$$V_0 \sim [1, \infty],$$

in which we can insure the emergences of the properties mentioned above, e.g. the small-world structure and $1/f^2$ noise. The ranges are wide, so the final network relies less on the initial state. As we can see, both of the structural and functional properties could emerge for a quite large range of parameters. For instance, the link probability p of the random network in our model can be taken between 0.06 and 1, i.e., which means the properties of emergence of small world structure and $1/f^2$ noise can exist even from sparsely to fully connected networks. This is strong evidence that the emergent properties are the results of inner co-evolutionary rules instead of fine-tuning of parameters.

Local rewiring rules, as well as the evolution rules of neuron states, are the key mechanisms in Bornholdt and APH models. These mechanisms are essential to produce the self-organized criticality (SOC) properties. Based on them, we developed a co-evolutionary model and applied it to an unbiased network. Unlike previous studies, the small-world structure is the result of evolution instead of the assumption beforehand. At the same time, SOC behaviors do not vanish. In other words, the coupled evolution process makes a dynamically stable structure and functional properties emerge simultaneously. This demonstrates our argument that they should emerge simultaneously in the same process since no evidence shows that their appearance should be one after another.

4. Conclusion

To summarize, we developed a neural network model with SOC features based on Bornholdt and APH models. Compared with other models, ours co-evolves from a random network with less biased structure. Emergence of SOC behaviors is observed during evolution, and the structural and functional features emerge simultaneously. Different from previous models, small-world topology emerged as a natural result of evolution, not an assumption *apriori*. The currents power spectrum exhibits power-law behavior, which is consistent with experimental results. Interestingly, the spectrum of average degree changes also exhibits power-law behavior, which is expected to be verified by future experiments.

Acknowledgments

This work is supported by NSFC of China under Grand no. 11175086.

References

[1] Chialvo D R 2010 *Nature Physics* **6** 744–750
[2] Beckers R, Deneubourg J L, Goss S and Pasteels J 1990 *Insectes Sociaux* **37** 258–267
[3] Rauch E M, Millonas M M and Chialvo D R 1995 *Physics Letters A* **207**(3) 185–193
[4] Beekman M, Sumpter D J T and Ratnieks F L W 2001 *Proceedings of The National Academy of Sciences* **98**(17) 9703–9706
[5] Clar S, Drossel B and Schwabl F 1996 *Journal of Physics-Condensed Matter* **8**(37) 6803–6824
[6] Peters O and Neelin J D 2006 *Nature Physics* **2**(6) 393–396
[7] Peters O, Hertlein C and Christensen K 2001 *Physcal Review Letters* **88**(1) 018701
[8] Peters O and Christensen K 2002 *Physical Review E* **66**(3) 036120
[9] Cavagna A, Cimarelli A, Giardina I, Parisi G, Santagati R, Stefanini F and Viale M 2010 *Proceedings of The National Academy of Sciences* **107**(26) 11865–11870
[10] Rybarsch M and Bornholdt S 2014 *PloS One* **9** e93090
[11] Kello C T 2013 *Psychological Review* **120** 230
[12] Gal A and Marom S 2013 *Physical Review E* **88** 062717
[13] Cowan J, Neuman J, Kiewiet B and van Drongelen W 2013 *Journal of Statistical Mechanics: Theory and Experiment* **2013** P04030
[14] Levina A, Herrmann J M and Geisel T 2007 *Nature Physics* **3** 857–860
[15] Millman D, Mihalas S, Kirkwood A and Niebur E 2010 *Nature Physics* **6** 801–805
[16] Rubinov M, Sporns O, Thivierge J P and Breakspear M 2011 *PLoS Computational Biology* **7** e1002038
[17] Touboul J and Destexhe A 2010 *PloS One* **5** e8982
[18] Bak P 1997 *How nature works* (Oxford university press Oxford)
[19] Bak P and Paczuski M 1995 *Proceedings of the National Academy of Sciences* **92** 6689–6696
[20] Bak P, Tang C and Wiesenfeld K 1987 *Physical Review Letters* **59**(4) 381–384
[21] Bak P and Chialvo D R 2001 *Physical Review E* **63**(3) 031912
[22] Magnasco M O, Piro O and Cecchi G A 2009 *Physical Review Letters* **102** 258102
[23] Beggs J M and Plenz D 2004 *Journal of Neuroscience* **24**(22) 5216–5229
[24] Bienenstock E and Lehmann D 1998 *Advances in Complex Systems* **1** 361–384
[25] Uehara T, Yamasaki T, Okamoto T, Koike T, Kan S, Miyauchi S, Kira J i and Tobimatsu S 2014 *Cerebral Cortex* **24** 1529–1539
[26] Koziol L F, Barker L A, Joyce A W and Hrin S 2014 *Applied Neuropsychology: Child* **3** 245–252
[27] Bornholdt S and R?hl T 2003 *Physical Review E* **67** 66118
[28] Hebb D O 1949 *The organization of behavior; a neuropsychological theory*
[29] De Arcangelis L, Perrone-Capano C and Herrmann H 2006 *Physical Review Letters* **96** 28107
[30] Bak P, Tang C and Wiesenfeld K 1987 *Physical Review Letters* **59** 381
[31] Rabinovich M, Varona P, Selverston A and Abarbanel H 2006 *Reviews of Modern Physics* **78** 1213–1265
[32] Rabinovich M, Volkovskii A, Lecanda P, Huerta R, Abarbanel H D I and Laurent G 2001 *Physical Review Letters* **87** 068102
[33] Abraham W C, Mason-Parker S E, Bear M F, Webb S and Tate W P 2001 *Proceedings of the National Academy of Sciences of the United States of America* **98** 10924–10929

[34] Hutchins J and Barger S 1998 *The Anatomical Record* **253** 79
[35] Johnson Jr E, Chang J, Koike T and Martin D 1989 *Neurobiology of Aging* **10** 549
[36] Kanazawa I 2001 *Trends in Molecular Medicine* **7** 339–344
[37] Livet J, Weissman T A, Kang H N, Draft R W, Lu J, Bennis R A, Sanes J R and Lichtman J W 2007 *Nature* **450** 56
[38] Hagmann P, Cammoun L, Gigandet X, Meuli R and Honey C 2008 *PLoS Biol* **6** 15
[39] Freeman W, Rogers L, Holmes M and Silbergeld D 2000 *Journal of Neuroscience Methods* **95** 111–121

The dependence of graph energy on network structure

Y Zhu[1,2,3], W Li[2,4], Q A Wang[1,3], X Cai[2]

[1] ISMANS, LUNAM Universitė, 44 av. Bartholdi, 72000 Le Mans, France
[2] Complexity Science Center & Institute of Particle Physics, Central China Normal University, Wuhan, 430079, China
[3] IMMM, UMR CNRS 6283, Universitė du Maine, 72085 Le Mans, France
[4] Max-Planck Institute for Mathematics in the Sciences, Inselst. 22, 04103 Leipzig, Germany.

E-mail: zhu@ismans.fr

Abstract. In this paper, we aim at investigating how the energy of a graph depends upon its underlying topological structure for regular and sparse scale free networks. Firstly, the spectra and energies of some simple regular graphs are calculated exactly and an exact expression is derived for the eigenvalues of adjacency matrix of regular graphs with degree k being given by $k = 2^a$ $(a = 1, 2, 3, \ldots)$. It is also found that a graph with k being about $0.8N$ owns the largest energy for the regular graphs with the same size and the same generating method used in this paper. Furthermore, we investigate the energy of sparse scale-free networks with different average degree $< k >$ and degree distribution exponent γ. While γ is specified, the energy is a power-law function of $< k >$ with exponent being about 0.5. And while $< k >$ is fixed, energy will be obviously proportional to γ. Otherwise, we also find that the energy is a power-law function of the variance of degree sequence with exponent weakly depending on the size of network. Interestingly, while both $< k >$ and γ are specified, there will be a terrific linear fit to the relationship between energy and the size of system.

1. Introduction

The graph energy E, defined as the sum of the absolute values of eigenvalues of adjacency matrix, is coming from chemistry where it used to approximate the total π-electron energy of molecules[1]. This definition form was first stated in the 1970s[2] and later proposed in the book[3] and paper[4]. Researches related to the graph energy can be traced back to the 1930s when a German scholar Erich Hückel proposed a method to approximately solve the Schrödinger equation of a class of organic molecules[5, 6, 7]. Erich Hückel obtained an expression for the total energy of all π-electrons. According to this expression, I. Gutman calculated that the energy is equal to two times the sum of the positive eigenvalues. By considering the symmetry of eigenvalues of all graphs, he changed the expression of graph energy to the sum of absolute values of all eigenvalues which is just formal and does not have a clear physical meaning.

For the classification of graphs according to the energy, a finite and undirected graph G with N nodes is denoted to be "hyper-energetic" if its energy is larger than $2(N-1)$[8], which corresponds to the energy of a complete graph on N vertices. And if the energy of a connected graph G is less than N(which relates to the graph generated by $N/2$ isolated edges or $N/4$ isolated quadrangles), then the graph will be labeled as "hypo-energetic" graph[9]. Another one that regards to the "hypo-energetic" graph is the strongly "hypo-energetic" graph with energy being less than $(N-1)$[10]. It is particulary noteworthy that the star graph with N nodes owns the lowest energy $2\sqrt{N-1}$ comparing with all the other connected graphs with the same vertices.

Until now, most researches on graph energy still remain at determining the bounds of energy for some particular graphs. For instance, J. H. Koolen and V. Moulton proved that the energy of graph G on N vertices was not larger than $\frac{N}{2}(1+\sqrt{N})$[11]. While both N and edges m are given, the graph energy will less than or equal to $\sqrt{2mN}$[12]. Other bounds of graph energy can be found in references [13, 14, 15, 16, 17, 18, 19, 20, 21]. However, even though some bounds of energy for graphs are well developed, relatively little is known regarding to the relationship between energy and the structure of graphs.

In this paper, we aim at exploring the dependence of graph energy on the graph's underlying structure. In the next section, we first calculate exactly the spectra and energies of some simple $k-$regular graphs. And the relation among energy and degree k for regular graphs is also studied in this part. Some distributions of energy regarding to some topological parameters of sparse scale free networks are displayed in section 3. And Section 4 presents our conclusions.

2. Graph spectra for some regular graphs

We begin with the investigation of some simple regular graphs, which are generated with all nodes being located in a ring and all edges being formed by nearest neighbor connections. In this part, we strictly calculated the eigenvalue spectra of adjacency matrix for some regular graphs with node's degree k being small. Furthermore, an exact function of eigenvalue spectra for some regular graphs with k being an exponential function of 2 is deduced.

Firstly, for the regular graph with $k=2$ (see Fig.1(a)), the eigenvalues of adjacency matrix can be expressed by

$$\{\lambda_i\}_0^{N-1} = 2\cos[\frac{2\pi i}{N}], \tag{1}$$

where N is the number of nodes. Therefore, according the definition of graph energy, we can strictly obtain the energy E:

$$E = \sum_{i=0}^{N-1} 2|\cos[\frac{2\pi i}{N}]| = \frac{4N}{\pi}. \tag{2}$$

And for the regular graph with $k=4$ (see Fig.1(b)), the eigenvalues are

$$\{\lambda_i\}_0^{N-1} = 4\cos[\frac{\pi i}{N}]\cos[\frac{3\pi i}{N}], \tag{3}$$

and graph energy is $\frac{3\sqrt{3}N}{\pi}$.

For the cases of $k=6$ and $k=8$, whose topological structures are showed in Fig.1(c) and Fig.1(d), the eigenvalues are

$$\{\lambda_i\}_0^{N-1} = 2\cos[\frac{6\pi i}{N}] + 4\cos[\frac{\pi i}{N}]\cos[\frac{3\pi i}{N}], \tag{4}$$

and

$$\{\lambda_i\}_0^{N-1} = 8\cos[\frac{\pi i}{N}]\cos[\frac{2\pi i}{N}]\cos[\frac{5\pi i}{N}]. \tag{5}$$

And the corresponding graph energies are $\frac{32\sqrt{2}-6\sqrt{3}}{6\pi}$ and $\frac{(5\sin[\frac{\pi}{5}]+25\sin[\frac{2\pi}{5}]-8)N}{3\pi}$ respectively.

So, for the specified regular graphs with $k = 2^a$, where a is an integer, we have derived an exact expression for the eigenvalues of adjacency matrix. That is

$$\{\lambda_i\}_0^{N-1} = k\cos[\frac{\pi i}{N}]\cos[\frac{2\pi i}{N}]\cos[\frac{4\pi i}{N}]\cos[\frac{8\pi i}{N}]\cdots\cos[\frac{(k/2+1)\pi i}{N}], \quad (6)$$

in which, the number of cos[...] function is a.

$(a) k = 2$ $(b) k = 4$ $(c) k = 6$ $(d) k = 8$

Figure 1. Topological structures of some regular graphs generated with all vertices being located in a ring and all edges being formed by nearest connections.

In 1978, I. Gutman put forward a conjecture to describe that a complete graph would have maximal energy[2]. However, soon thereafter, this conjecture was shown to be false by means of counterexamples[22]. So in this part, we will investigate whose energy will rank first for those regular graphs with the same generating method. Fig. 2(a) shows the distribution of graph energy E in the variation of k for different values of N. It can be obviously noticed that there exist a peak in the change curve of graph energy for each N. When the network is very sparse, graph energy E will be monotonically increasing with k. With slow increase of intensity, energy E will be still proportional to k but with appearing of some weak fluctuations. However, while the network is becoming highly dense, the graph energy will drop sharply. And the value of k that corresponds to the largest graph energy is about $0.8N$. Fig. 2(b) presents the normalized result for the increasing part of the left one. It displays that while the regular network is very sparse, the normalized graph energy follows a power-law function of k, which can be expressed by $E/E(k = 2) = 0.977k^{0.214}$. And the two parameters in this function are almost independent of the size of system. Particularly it is worth mentioning that the graph energy corresponding to complete graph ($k = N - 1$) or sub-complete graph ($k = N - 2$) is equal to $2k$.

3. Some results in sparse scale free networks

The correlation between energy and average degree, denoted by $< k >$, is also explored firstly in this part, as it is depicted in Fig. 3, which plots the normalized energy as a function of $< k >$ for two different situations with γ being 2.5 and 6.0 (γ is the exponent of degree distribution of network considered). The fitting line suggests a power law function between E and $< k >$, which can be expressed as follows.

$$E/E(< k >= 3) = 0.570 < k >^{0.502}, \quad (7)$$

in which, the three parameters are almost independent of the value of γ as long as $< k >$ is small.

Furthermore, the effect of degree exponent γ on graph energy is also studied. Plots presenting normalized graph energy distribution for different network sizes are shown in Fig. 4, which provides an intuitive perspective to realize the proportional relationship between γ and E when γ is not so large. After γ gets to 7, the plateau in the energy distribution is presented obviously, especially for the smaller systems. Plateau corresponds to the network with a regular structure. Apparently, for larger systems, the value of γ resulting in a regular structure network will be

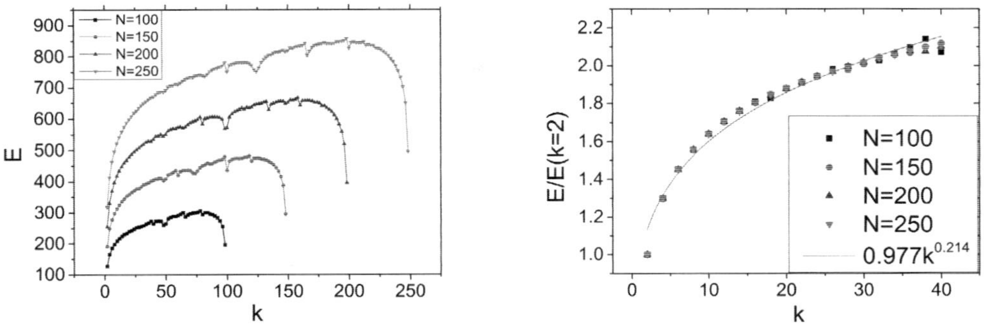

(*a*) (*b*)

Figure 2. (Color online) (*a*) shows the dependence of graph energy E on degree k for different values of N. (*b*) with a solid power law fitting line describes the normalized results for the increasing part of left plot without any fluctuations. It shows a power law correlation between E and k while k is small.

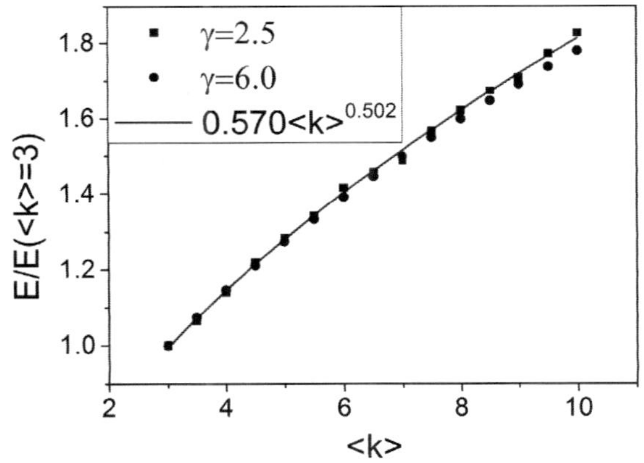

Figure 3. The normalized graph energy as a function of average degree $< k >$ for the cases of $\gamma = 2.5$(squares) and $\gamma = 6.0$(circles) with $N = 1000$. The solid line is a power law fitting line for the case of γ being 2.5.

larger. And on the whole, the dependence of graph energy on degree exponent γ will be stronger with the increasing of N.

Describing it differently, the correlations between graph energy and degree distribution can be interpreted by the relations between it and the variance of degree sequence. The variance, denoted by D, is defined as

$$D = \frac{1}{N}\sum_{i=1}^{N}(k_i - < k >)^2, \tag{8}$$

where k_i means the degree of ith agent.

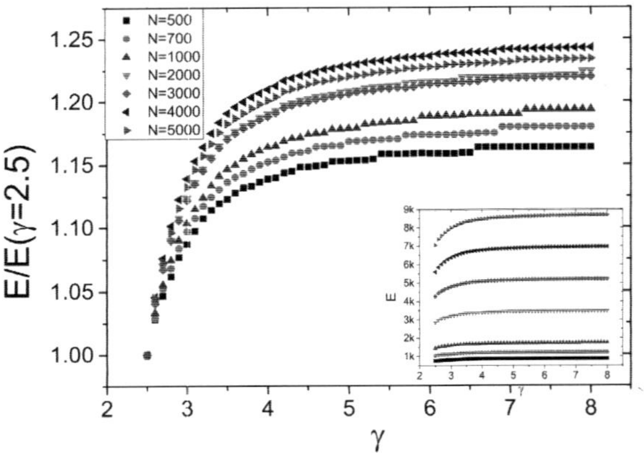

Figure 4. (Color online) The distribution of normalized graph energy with the variation in γ for some specified values of N. Inset shows the distributions of graph energy before being normalized.

To analyze the impacts of the variance of degree sequence on the variability of graph energy, we propose the relative variation of energy, labeled as δE. δE is defined by $(E - E_{max})/E_{max}$, where E_{max} denotes the graph energy with the condition of D being close to 0 which corresponds to a uniform degree distribution. Fig. 5 shows how the relative variation of graph energy relies on the degree variance for different values of N. This shape of energy distribution represents a power-law correlation between δE and D, that can be fitted by the following function.

$$\delta E = -yD^{\beta}. \tag{9}$$

Fig. 6 displays the dependence of parameters y and β in equation (9) on the size of system N. There exists a smooth power law fitting line to the relationship between parameter β and N, which can be described by $\beta = 1.251N^{-0.075}$. And according to Fig. 6(b), it can be obviously noticed that parameter y almost does not change with increasing of N but owns a fixed value close to 0.013. So, we can draw easily that the dependence of variability of graph energy on the variance of degree sequence will be weak with the increasing of N.

We have known about the linear relationship between graph energy and the size of system for the regular graphs (all nodes are located in a ring and all edges are generated by nearest connections) with specified degree k from theoretical view. Similarly, we study the distribution of graph energy over N with specified average degree $< k >$ and degree exponent γ, see Fig. 7. It indicates that the energy is also linearly increasing with N for each value of γ, that is

$$E \propto \alpha N, \tag{10}$$

and α will increase with γ while γ is small according to Fig. 7. And the overlap of the results under the conditions of $\gamma = 6.0$ and $\gamma = 8.0$ is simply on account of the uniform degree distribution (i.e, the corresponding network is regular) after γ gets to 6.0. With the increasing of γ from 6.0, the structure of network will hardly change.

4. Conclusions

We first calculated exactly the energy of regular graphs for node's degree k being small. And an exact formula for the eigenvalues spectra of regular graphs with $k = 2^a$ $(a = 1, 2, 3, \ldots)$ is

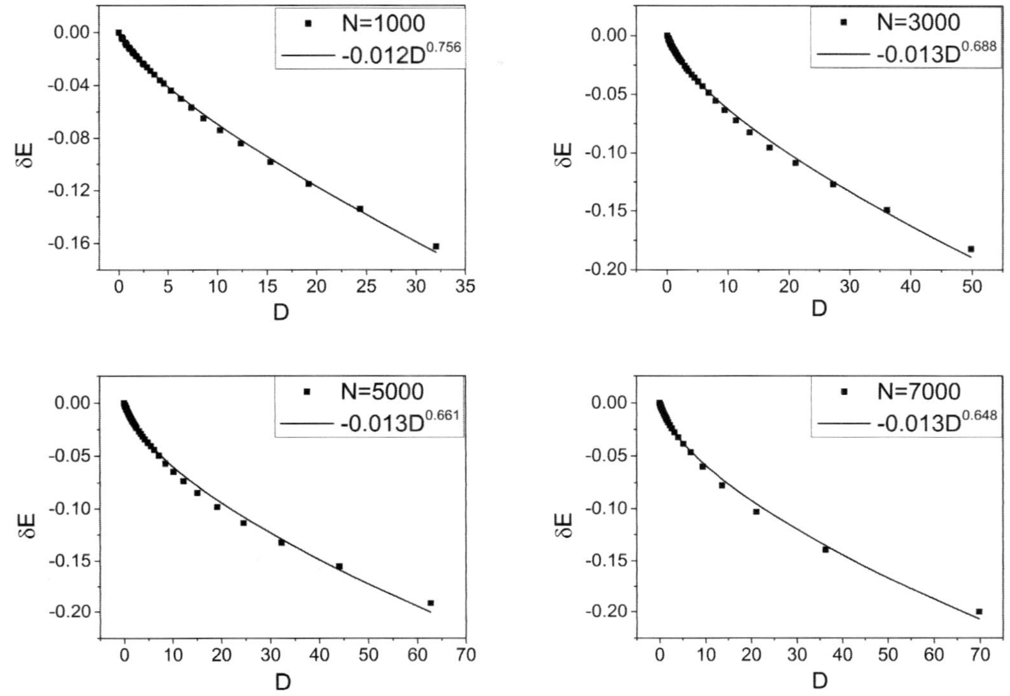

Figure 5. The relative variation δE of graph energy as a function of D for systems with different size and average degree $< k >$ being 4. The solid lines are obtained by power-law fitting.

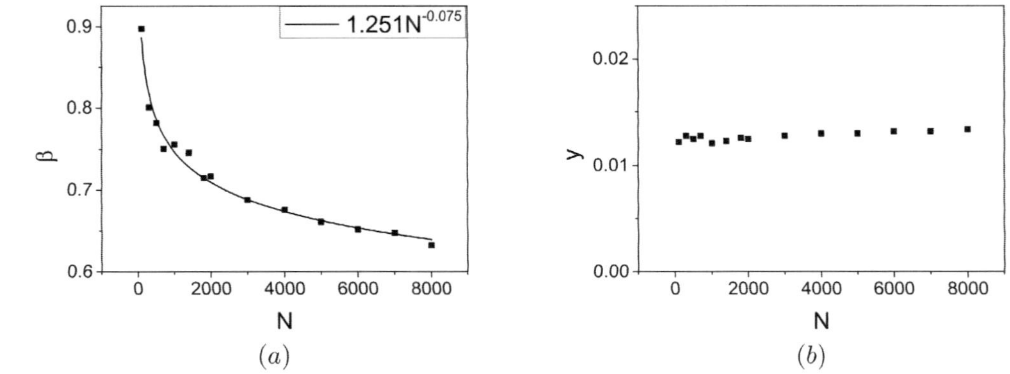

Figure 6. (a). The distribution of exponent β in equation (9) with the variation in N for average degree $< k >$ being 4. And the solid line is fitted by a power law function. (b). The dependence of another parameter y in equation (9) on N. As can be seen on the plot, parameter y has barely anything to do with N.

deduced. We also find that the graph with k being about $0.8N$ owns the largest energy for

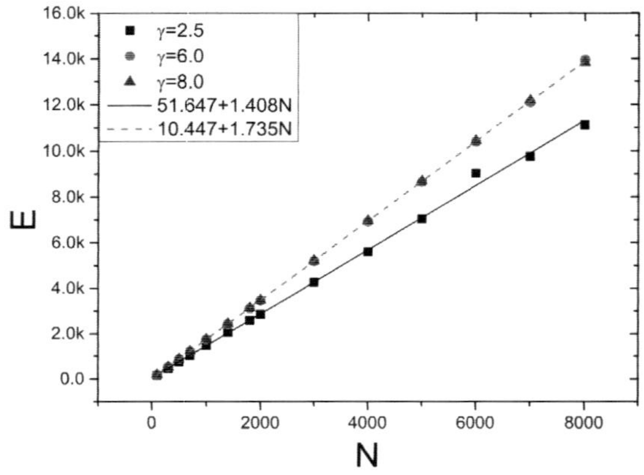

Figure 7. (Color online) The distributions of graph energy varying with system's size N with average degree $< k >= 4$. The squares correspond to the result of $\gamma = 2.5$. The circles and triangles correspond to the cases of $\gamma = 6.0$ and 8.0 respectively. And the dash and solid lines are linear fitting lines for the corresponding distributions of graph energy.

regular graphs (generated in such a way that all nodes are positioned on a ring and edges are formed by nearest neighbor connections) with specified N. Subsequently, we investigated the dependence of energy on the structure of graph, such as the average degree, the degree exponent and degree variance for sparse scale free networks. The results illustrate that graph energy is a power-law function of average degree $< k >$ when degree exponent γ is fixed and is also a power-law function of degree variance D while $< k >$ is specified. And the dependence of graph energy on degree exponent will be stronger with the increasing of N while the case is opposite for its dependence on degree variance. It is worth mentioning that the energy is always linearly increasing with the size of system in regular and scale free graphs.

Acknowledgments
This work was supported by China Scholarship under Grant No. 201406770035 and the Programme of Introducing Talents of Discipline to Universities under Grant No. B08033.

References

[1] Gutman I 1977 *Theor. Chim. Acta.* **45** 79-87
[2] Gutman I 1978 *Ber. Math. Statist. Sekt. Forsch. Graz.* **103** 22
[3] Gutman I 1986 *Mathematical Concepts in Organic Chemistry* (Berlin: Springer-Verlag)
[4] Gutman I 1987 *J. Math. Chem.* **1** 123-43
[5] Streitwieser A 1961 *Molecular Orbital Theory for Organic Chemists* (New York: Wiley)
[6] Yates K 1978 *Hückel Molecular Orbital Theory* (New York: Academic Press)
[7] Coulson C A, O'leary B and Mallion R B 1978 *Hückel Theory for Organic Chemists* (London: Academic Press)
[8] Gutman I 1999 *J. Serb. Chem. Soc.* **64** 199-205
[9] Gutman I 2007 *Indian J. Chem.* **46** 1733-36
[10] So W, Robbiano M, Abreu N M M and Gutman I 2010 *Lin. Algebra Appl.* **432** 2163-69
[11] Koolen J H 2001 *Adv. Appl. Math.* **26** 47-52
[12] McClelland B J 1971 *J. Chem. Phys.* **54** 640-43
[13] Gutman I 1996 *Coll. Sci. Papers Fac. Sci. Kragujevac* **18** 75-82
[14] Koolen J H and Moulton V 2003 *Graphs and Combinatorics.* **19** 131-35

[15] Rada J and Tineo A 2004 *J. Math. Anal. Appl.* **289** 446-55
[16] Zhou B 2004 *Kragujevac J. Sci.* **26** 5-12
[17] Balakrishnan R 2004 *Lin. Algebra Appl.* **387** 287-95
[18] Pena J A, Mendoza L and Rada J 2005 *Discr. Math.* **302** 77-84
[19] Zhou B, Gutman I, Pena J A, Rada J and Mendoza L 2007 *MATCH Commun. Math. Comput. Chem.* **57** 183-91
[20] Gutman I, Firoozabadi S Z, Antonio J and Rada J 2007 *Math. Comput. Chem.* **57** 435-42
[21] Li X, Shi Y T and Gutman I 2012 *Graph Energy* (New York: Springer)
[22] Cvetković D and Gutman I 1986 *J. Comput. Chem.* **7** 640-44

AUTHOR INDEX